Natural Resources and the Environment Series
Volume 10

Integrated Physical, Socio-economic and Environmental Planning

NATURAL RESOURCES AND THE ENVIRONMENT Series

Volume 1: ENVIRONMENTAL IMPACTS OF PRODUCTION AND USE OF ENERGY
Essam El-Hinnawi, United Nations Environment Programme.

Volume 2: RENEWABLE NATURAL RESOURCES AND THE ENVIRONMENT: PRESSING PROBLEMS IN THE DEVELOPING WORLD
Kenneth Ruddle and Walther Manshard. United Nations University.

Volume 3: ASSESSING TROPICAL FOREST LANDS: THEIR SUITABILITY FOR SUSTAINABLE USES
Richard Carpenter, Editor, East-West Center

Volume 4: FUELWOOD AND RURAL ENERGY PRODUCTION AND SUPPLY IN THE HUMID TROPICS
W. B. Morgan and R. P. Moss, United Nations University.

Volume 5: ECONOMIC APPROACHES TO NATURAL RESOURCE AND ENVIRONMENTAL QUALITY ANALYSIS
Maynard M. Hufschmidt and E. Hyman, Editors, East-West Center.

Volume 6: RENEWABLE SOURCES OF ENERGY AND THE ENVIRONMENT
Essam El-Hinnawi and Asit K. Biswas, Editors.

Volume 7: GLOBAL ENVIRONMENTAL ISSUES
United Nations Environment Programme, *Essam El-Hinnawi and Manzur-Ul-Haque Hashmi,* Editors.

Volume 8: THE WORLD ENVIRONMENT 1972-1982
Martin Holdgate, Mohammed Kassas and Gilbert White, Editors; *Essam El-Hinnawi,* Study Director, United Nations Environment Programme.

Volume 9: ENERGY ALTERNATIVES IN LATIN AMERICA
Francisco Szekely, Editor. United Nations Environment Programme.

Volume 10: INTEGRATED PHYSICAL SOCIO-ECONOMIC AND ENVIRONMENTAL PLANNING
Yusuf J. Ahmad and Frank G. Müller, Editors. United Nations Environment Programme.

Volume 11: THIRD WORLD AND THE ENVIRONMENT
Essam El-Hinnawi and Asit K. Biswas, Editors.

Volume 12: DEVELOPMENT WITHOUT DESTRUCTION: EVOLVING ENVIRONMENTAL PERCEPTIONS
Mostafa K. Tolba.

Integrated Physical, Socio-economic and Environmental Planning

YUSUF J. AHMAD
and
FRANK G. MÜLLER, *editors*

Published for the UNITED NATIONS ENVIRONMENT PROGRAMME
by TYCOOLY INTERNATIONAL PUBLISHING LIMITED
DUBLIN

Tycooly International Publishing Limited
6, Crofton Terrace,
Dun Laoghaire,
Co. Dublin, Ireland
Telex 30547 SHCN EI

First edition 1982

Typeset by Africa Science International Publishing Ltd., Nairobi, Kenya and Printed by Cahill Printers Limitted Dublin Ireland.

ISBN 0 907567 18 5 Hardback.

ISBN 0 907567 19 3 Softcover.

CONTENTS

		Page
Preface		vii
Foreword		ix
Introduction		xii

CONCEPTS

Chapter 1	Current work in the field of Integrated Physical, Socio-economic and Environmental Planning and the United Nations Economic Commission for Europe: A Survey *K. Leonhardt*	1
Chapter 2	Environmental Management: Issues of an Integrated Planning Approach *F. G. Müller*	15
Chapter 3	Institutional Aspects of Integrated Planning Processes *Klaus Zimmermann*	35
Chapter 4	Interactive Learning Models for Integrated Planning *Peter Nijkamp and Jaap Spronk*	53

COUNTRY EXPERIENCES

Chapter 5 Public Participation in Environmental Planning
Dorothy Nelkin 71

Chapter 6 Integrated Physical Planning in Sweden
Per Wramner 93

Chapter 7 The Political Economy of Pollution Control in the United States
Paul B. Downing 109

Chapter 8 The Integrated Planning System and Economic Reform in Poland
Bohdan Gruchman 137

Chapter 9 Experiences of Environment Conservation in the Uzbek SSR
Igor V. Chirgadze 151

Chapter 10 The Effects of Socio-economic Activities on the Environment in the framework of Integrated Physical Planning: Some French Experiences
Francis Wasserman 155

Chapter 11 Transboundary Environmental Issues in the US - Mexico Borderlands
Niles Hansen 167

CONCLUSIONS

Chapter 12 Conclusions and Recommendations 191

List of participants 197

Preface

THE incorporation of environmental concerns in social decision-making is a necessary but complex task. Integrated physical, socio-economic and environmental planning has been developed in a number of countries, amongst them Sweden, France, Japan and the centrally planned economies of Eastern Europe, as an effective instrument for the purpose.

It is not an easy instrument to develop or to apply. In order to fashion it, that is to say, in order to give the ideas and concepts behind the approach a certain level of operational content, a number of essential pre-conditions must be met. Firstly, the data and information base for preparing such an integrated plan must be available. Secondly, it will be necessary to have the administrative machinery, at the local and regional levels, to assist, monitor and implement the plans when made. Finally, there is the requirement of a threshold level of public awareness and participation, starting from the grass-roots, in decision-making. Most developing countries might find these conditions difficult to meet, at least comprehensively, in the beginning. There is, however, no reason why preliminary steps towards identifying and examining the basic ground rules and prospects for such integrated planning could not be initiated with advantage without undue loss of time.

Sound environmental management is essentially a problem of allocation of scarce resources between different possible uses. Environmental considerations join hands with economic criteria in demanding that such allocation be as efficient as possible. Techniques of integrated physical planning attempt to realise that objective. There is much to be currently learned both from the theoretical frameworks that have been prepared by experts and the practical lessons in the field in different countries.

It is the intention of the United Nations Environment Programme (UNEP), at the request of its Governing Council, to examine theoretical and empirical evidence to distinguish what is feasible for developing countries at present and to prepare manuals and guidelines for action. This volume is an examination by a number of well-known international experts of the problems and issues involved in integrated physical planning and is the first step in that direction.

Mostafa K. Tolba
Executive Director
United Nations Environment Programme

Nairobi, January 1982

Foreword

AT ITS seventh session, the Governing Council of the United Nations Environment Programme (UNEP) adopted Decision 7/7 on *Environment and Development Activities,* Part A of which reads as follows:

> "Recognizing the importance of the development taking place in the developing countries and the implications of action taken both in the urban and in the rural environment;
>
> Requests the Executive Director, in co-operation with the specialized agencies of the United Nations system, to promote and contribute to the setting up of studies of the environmental impact of socio-economic activities in the context of an integrated physical planning policy."

The Decision reflected the view expressed by several developing countries during the debate in the Council that it had become a matter of some urgency and importance for them to incorporate environmental considerations in development planning and a growing consensus that long-term and sustainable development could only be ensured through sound environmental management had emerged. The objective they sought to realize was to facilitate economic development, while avoiding or minimizing the concomitant environmental damage.

Since then, UNEP has been attempting to examine the statistical data base available in several developing countries as well as the institutional machinery, existing at the local and regional levels, which could be harnessed in the fashioning of integrated physical, socio-economic and environmental planning and implementation procedures.

We are living in a period of rapid change, faced with an unstable economic situation and an increasingly complex society. Scientific and technological developments have opened up unprecedented opportunities as well as new dangers and problems. New social objectives need to be identified and defined that will take into account the impact of technological innovations and developments, improved living standards for an increasing number of people, with consequential demand for space and durable consumer goods, and the need for the population to adapt to structural changes, both in relation to the availability of resources and the use of increased leisure time.

In this situation we require, as never before, a clear understanding of the interrelationship between population, resources, environment and development. New ways have to be sought to overcome the complexities which 'sectoral' planning cannot meet. A more 'global', although selective and flexible, planning process needs to be employed. Such a planning process would be an integrated and continuous one. It could not be a single-focus activity.

What is sought is a tool that is more refined than comprehensive planning (concentrating on key parameters and levers of social and economic policy rather than on an indiscriminate collection of data and issues) and more flexible than co-ordinated planning (no rigid frameworks are sought to be imposed). Such a tool would enable growing structural problems in the economy to be met. Structural problems could be of national, regional or local import and must be met in the context of resources and energy constraints. Integrated planning is based on the explicit recognition of the interdependence between socio-economic dynamics and the physical conditions within which they operate.

But such a planning process, which has a direct impact on public well-being and, therefore, on public opinion, is by its nature a difficult exercise both in terms of construction and implementation. This development of a coherent assessment and management process would appear to depend upon the presence of a number of contributing elements, including:

1. the availability and handling of data and information on resource levels and environmental concerns generally;
2. the presence of mechanisms and procedures for carrying out environmental impact assessments of different types of development activities;
3. monitoring of an evolving process so that recommended policies, priorities and techniques of planning could be examined and changed, based on a systematic feedback of experience;
4. flexibility in the planning approach;
5. the identification of alternative scenarios in terms of exploitation of limited natural resources, the impact on the employment situation, the development of regional and local potential, the quality of life in human settlements, etc.;
6. the development of alternative scenarios taking the major parameters into account;
7. popular participation at different levels of decision-making;
8. the availability of minimal institutional mechanisms and modalities at the regional and local levels; and,

9. above all, a conscious policy attempt to strike a balance between environmental, social, political and economic concerns.

The practical problem, thus, arises from the fact that although it is easy to recognize the need for integrated planning, it is more difficult to devise policies, institutions and systems to ensure that it is carried out on a continuing basis. It is not surprising that integrated plannng is still very much a process of learning from experience. It must be based firmly on what is politically and administratively feasible. There could be no single approach to integrated planning, though many elements and techniques are of common interest to governments at different levels.

On the other hand, it is becoming increasingly urgent to incorporate environmental considerations into the decision-making process for which integrated planning provides a useful vehicle. There are, for instance, a number of major economic, particularly investment, and social decisions pending for the developing countries—decisions that would have long-term effects, on a significant scale, with a high element of uncertainty and risk—in such areas as the future of urban settlements, including transport and communications, food production and energy utilization, industrial growth and the like. Complex and wide ranging decisions that must be made in the field of energy alone—new sources of renewable energy that must be tapped and new sites for power stations that must be determined—are truly daunting.

The developing countries are, thus, already faced in a concrete manner with the challenges of integrated planning, that is to say, of carrying out a series of simultaneous and sequential evaluation of data and information and of rapidly changing interrelationships to secure desired objectives. In order to succeed in this difficult task, it is necessary (a) to adopt anticipatory policies for the integration of environmental considerations in economic and social decision-making; (b) to accept improvement in the quality of life as a basic element of social policy; and (c) to encourage popular participation in the resolution of conflicts and trade-offs between different options available.

It is evident that these tasks, involving techniques of thinking and attitudes of mind, cannot be accomplished alone, either by UNEP at the global level or the environmental machinery of the developing countries at the national level. Governmental machinery, industrial structures and academic expertise must all co-operate to identify a new way of thinking about systemic relationships and determine means to effectively manage the processes of social growth and change through fluid and adaptive mechanisms.

Yusuf J. Ahmad
Director for Special Assignments
United Nations Environment Programme

Nairobi, January 1982

Introduction

THE purpose of this book is to examine a wide range of issues on integrated physical planning. In many countries, the interest in sound and comprehensive environmental management policies and planning methods has been growing rapidly in the last decade. In retrospect, it seems that many traditional approaches to society's demands no longer hold. The worldwide slowing down of economic growth questions the importance of current environmental policies and management practice. In a re-orientated discussion, the aim is now to make environmental protection compatible with other socio-economic goals.

In this respect, decision-makers and planners, confronted with the present problems arising from energy strategies, rising unemployment and price levels, are now facing planning processes which have become increasingly complex, as many policy objectives have to be pursued simultaneously. These processes are also complicated by the fact that there seems to be no agreement among the various actors concerning the priorities of the various objectives. Consequently, there is a basic need to understand the decision-making and planning processes at work, the kind of policy decisions that are desirable and feasible, and the consequences of these decisions and their possible alternatives. In this context, two observations can be made. Firstly, with respect to the use of concepts and models of environmental analysis, there are several deficiencies in the following areas: the range of applicability (for example, in a political-institutional setting); the limitations of partial analyses (for example, an irrigation study without an income-redistribution analysis); and the restrictions arising from the available data and the transferability of results among countries and regions. Secondly, in the field of environmental management policy, several problems are encountered: the provision of relevant policy information; the assessment of the impacts of socio-economic activities on the environment; the determination of the cost-effectiveness of environmental policies; and the lack of public consensus on environmental issues.

Consequently, an integrated view of environmental management requires multi-disciplinary knowledge and an inter-disciplinary approach to develop public policy programmes that enhance and achieve the politically desired quality of life. Many of these problems have been discussed at the *Ad Hoc* Expert Meeting on Integrated Physical, Socio-economic and Environmental Planning, which was held at the Industry and Environment Office of the United Nations Environment Programme (UNEP) in Paris, France, in September 1981.

This volume is divided into two parts. The first section includes contributions dealing with several conceptual issues of integrated planning, in a fairly general way. In a sense, they may be regarded as a descriptive and analytical frame of reference for the case studies which follows in section II.

Leonhardt presents a survey of the activities on integrated physical, socio-economic and environmental planning, undertaken by the United Nations Economic Commission for Europe (ECE). Several subsidiary bodies of the ECE are involved in assessing the environmental impacts of economic and technological development. The activities of various ECE bodies have contributed to the increasing awareness that environmental problems are of a comprehensive nature and have to be included in the planning and decision-making process. Thus, sectorial planning has been developed into an integrated planning approach.

Müller analyses a broad range of issues concerning environmental management, based on an integrated planning approach. This complex management task requires the environmental decision-maker to recognize the key factors which constrain his activities, namely social, technical, political, institutional, ecological and economic factors. After discussing the problems of defining and setting social goals and evaluating criteria for trade-off analysis, the main emphasis of this article is focused on the integration of environmental impact statements with cost-benefit analysis. The proposed strategy requires an explicit statement of the economic desirability of the projects, in the form of an EIS analysis. This suggested procedure thus requires the decision-makers and planners to arrange and evaluate the information to make an explicit judgement about the economic and environmental values involved. However, one must realize that, independent of the planning approach chosen, all decisions ultimately have to be sanctioned by the holders of political power.

The importance of institutional structures in integrated planning methods is evaluated by Zimmermann. Possible criteria for defining institutions are reviewed and then evaluated in terms of their usefulness for realizing and implementing integrated planning approaches—apparently two quite different problems. Since the explanatory power and the operationalization of the suggested criteria are hardly sufficient for designing an optimal institutional structure, he suggests a set of general rules and competencies instead, as preconditions for centralized, and/or decentralized structures as a prerequisite for successful implementation of integrated planning. However, in the final analysis, he concludes that the institutional question of the assignment of political responsibilities and powers represents a genuine political problem and cannot be solved with the help of scientific methods alone.

In Chapter 4, Nijkamp and Spronk attempt to design a methodological background for policy analysis. Society's demands require policy analyses to be multidimensional in nature in order to accommodate a wide variety of social interests, decision groups and policy structures. Despite the fact that programming models, in particular operation research methods, have played an important part in solving economic and environmental problems, such models have been criticised for their restrictive assumptions underlying such advanced formalized techniques. Thus, a broader view on policy analysis is required, based on an interactive framework for evaluating alternative policy options. From their formalized model of policy analysis and available experiences, the authors draw the conclusion that interactive policy models can be regarded as an important decision aid. Furthermore, it should be noted that the main purpose of the models is not to achieve 'optimal' solutions but rather to serve as essential learning tools, both for decision-makers and analysts. This is particularly true for cases where, very often, no exact goal definitions exist and, in this context, interactive policy models can provide assistance to define the decision-makers' objectives.

Nelkin discusses the problems of public participation in the process of environmental planning in the USA. She evaluates the arguments brought forward in favour of and against public participation in the decision-making process. Although admittedly time-consuming, it is assumed that public input into governmental decisions will contribute substantively to the quality of decision-making in general and will enhance environmental protection in particular. Thus, active and broadly supported public participation will assure that governmental policies reflect public preferences. Opposed to widespread public participation are those who are charged with administrative responsibility. They resent public participation because they consider it as interfering with their decision-making process and, thus, as an obstacle to implementing policies efficiently. The position towards public participation remains ambivalent and inconclusive, due to insufficient empirical experiences. As a concept, public participation represents a democratic right to information but as a procedure, it may lend itself to inefficiency and obstruction and lead to delays in technological progress and economic growth.

The integrated physical planning process and its institutional setting in Sweden is described by Wramner. The author stresses that, traditionally, physical planning is principally a municipal responsibility but that the national government has been playing a more active part in the planning process during the last decade. The emerging, national physical planning has completed its introductory phase and must now be integrated with the day-to-day planning process, particularly at the municipal level. Despite different beneficial consequences of national physical planning, the author stresses the problems of its implementation and considers wider decentralization and increased pubic participation as desirable.

The interactions between politics and economics in the integrated planning process, in the context of the United States, are analysed by Downing. The institutional incentives and objectives of the various actors (political entrepreneurs, bureaucrats, firm managers, citizen-interest groups and the public sector) occupying public and private positions lead to a complex set of mutually beneficial exchanges. During this political bargaining process, benefits accrue continuously to some and costs are imposed on others, whilst timing and incidence of benefits and costs may vary among different economic sector. The influence of any one group in this political bargaining process will differ among issues and for regions.

The adoption and implementation of pollution laws (for example the US Clean Air Act) are the result of a complex bargaining process. The distribution of property rights to the environment is central in this process for determining who is paying for pollution control. The economics and politics of environmental protection in the US have resulted in a shared ownership between polluters and pollutees, much like that in other countries. However, the author argues that while it appears that the pollutees have more rights under the US environmental protection laws compared to other countries, in reality they have less.

Gruchman looks at the Polish planning system including its physical, socio-economic and environmental aspects. The present shape of this integrated planning system was predominantly influenced by the 1961 law of 'spatial planning', which related physical planning to long-term socio-economic planning. The main characteristic of the system, as it evolved over the last two decades, is the national spatial plan which represents an integral part of the long-term, national development plan. The national spatial plan deals with all major physical, socio-economic and environmental development problems over national space.

The major shortcomings of the Polish integrated planning system are, to a considerable extent, part of a more general weakness of the entire planning system as it operated until very recently, namely insufficient links and co-ordination between the national spatial plans and its subordinate counterparts. The on-going discussion about the Polish economic system, through possible increased public participation in decision-making and the introduction of additional elements of a market economy.

Chirgadze succinctly reviews how environmental planning and protection measures have been integrated into development plans in the Uzbek Soviet Socialist Republic and how these plans have been set up in the context of a socialist economy.

The changing perception of regional development and planning in France is discussed by Wasserman, from a predominantly economic approach to the inclusion of environmental problems in the planning process. At present, physical integrated planning is influenced by the evolution of macro-economic and sectorial planning, as well as the consideration of environmental factors.

Hansen describes the recent situation of environmental quality along the Mexico-USA border. This area represents a unique interface between an advanced industrial nation and a nation that is progressing economically but that still shares many of the values and attitudes of Third World countries. Economic development issues have dominated environmental concerns on the Mexico side as well as in the Texas borderlands. In the more economically advanced borderlands of the western USA, environmental quality has received much more attention but unilateral actions in this regard have caused conflicts with the more development-oriented Mexican side. The need for better intemanagement of environmental resources along the border has been frequently suggested. But the author concludes that international agreements between national agencies are not likely to produce tangible results, as long as border communities have different values and, consequently, do not demand action.

This volume provides an extensive set of new themes and methods for analyzing environmental management problems. Instead of fairly narrow, mono-disciplinarian approaches, there is an increasing need for an integrated, inter-disciplinarian way of thinking, in which all economic, social and political sciences, ecology and social engineering constitute the foundation for a more comprehensive and integrated view of environmental management and planning

The present contributions show indeed that there are new ways of approaching current environmental issues but that there is no unambiguous method for each kind of environmental problem. We trust that some useful insight into the nature of integrated physical, socio-economic and environmental planning has been given.

Frank G. Müller

Nairobi, January 1982 United Nations Environment Programme

CHAPTER 1

Current Work in the Field of Integrated Physical, Socio-economic and Environmental Planning and the United Nations Economic Commission for Europe: A Survey

K. Leonhardt

BACKGROUND

SINCE 1950, population growth in Europe and North America has been coupled with significant improvements in material living standards, as the essential requirements were generally being met in sufficient quantity to satisfy the basic needs of the majority of people. Moreover, partly due to the impact of technological innovations and developments, improved living standards have given rise to an increasing demand for space and durable consumer goods. Such growth has brought about almost worldwide economic and social changes and, with these, problems have arisen; for example, employment and industrial problems of a structural nature have become major issues for a number of nations. There are many indications that such structural changes in the economy are of a permanent nature, to which the people of the region will have to adapt, in relation both to the availability of resources and the use of increased leisure time.

While the pace of change has varied across the ECE region, there is similarity in a special context. For example, settlements, flexible to a limited extent in their physical structure, exist in a dynamic environment of changing social needs and economic opportunities. There is a growing recognition of the inter-dependence between social and economic systems and the physical setting of land, building and infrastructure within which they operate.

Paradoxically, growth and change, while solving old problems, have created new ones. There has been widespread and growing concern about the nature of modern life and the mechanisms and institutions which brought about change. Among other concerns, questions have been raised about the pressure being exerted on scarce environmental and other resources. In addition, public participation in the decision-making process has been steadily increasing and has found greater expression.

The response of the planning process to the factors of growth and change, which previously seemed clear and unequivocal, now appear far more diverse. The context within which planning takes place is a fluid one. Dynamism within society makes it necessary for planning to be sufficiently adaptive to cope with changing values. That this is happening is reflected by one of the more noticeable features over the last three decades: the rapid expansion of government intervention and activity as a reaction both to changing values in society and the need to control the ever-increasing scale of development. A further point is that, with the success in fulfilling the essential requirements of people, the emphasis of planning has changed. A far greater interest is now apparent in improving the quality of life, rather than concentrating solely on quantitive gains. This has placed in relief the problem of measuring achievement in planning.

The traditional approach, of each activity being planned independently, has revealed too many deficiencies. In the mid-1960s, a much clearer recognition was gained of the complexities linked to social change inherent in planning during a period of economic growth, although the appreciation of the existence of intricate social and economic systems made the planning activity much more daunting. New ways had to be found to manage effectively the process of change. 'Integrated planning' has, thus, been suggested as an explicit attempt to deal with this.

It has been recognized that the problems and opportunities confronted by planning have diverse origins. Policies designed to tackle one issue also affect the scope of other, interrelated issues. Thus, planning can no longer be seen primarily as a single-focus activity. It has gradually become refocused by acknowledging explicitly the interplay of the social, economic and environmental forces at work in society. A picture of the state of planning in the ECE region at any moment would be but a single 'frame' of a film-strip sequence. Planning has been developing in such a way as to begin to come to grips with the complexity of the modern world. Theoretically and, increasingly so, practically, planning has been accepted as an integrated and continuous process. The development of new methodologies and analytical techniques have made it possible for inter-relationships to be explicitly identified. Nevertheless, there has been a tendency for each level or sector of planning to develop its own area of expertise and corporate identity, with all the consequent implications for problems which fall between disciplines or jurisdictional boundaries. It has, therefore, become clear that many problems cannot be dealt with effectively without an integrated approach.

ACTIVITIES OF ECE BODIES RELATED TO ENVIRONMENT-IMPACT ASSESSMENT

The assessment of environmental implications of specific economic and technological developments is proceeding through the programmes of several Principal Subsidiary Bodies of the Economic Commission for Europe (ECE). These considerations are taken into account in the work programme of a subsidiary body on an *ad hoc* basis, as the likelihood of an adverse impact on the quality of the environment becomes apparent. Activities, relative to the assessment of environmental impacts of economic and technological development within the context of ECE work, assumed formal and deliberate proportions with the implementation of the Final Act of the Conference on Security and Co-operation in Europe. In particular, the Parties to the Act agreed: "To advocate that within the framework of the United Nations Economic Commission for Europe, a study be carried out of procedures and relevant experience relating to the activities of Governments, in developing the capabilities of their countries to predict adequately environmental consequences of economic activities and technological development."

At its last session, in April 1981, the Commission requested "its subsidiary bodies to continue to take into account, for their work, the pertinent provisions of the Final Act, which call for multilateral implementation within the framework of the ECE" (Commission Resolution 1 (XXXVI), operative para. 2).

At their seventeenth session, held in November 1980, the Senior Economic Advisers to ECE Governments emphazised the need for a deeper and broader economic analysis of the necessary orientation of future investment patterns toward environmental purposes and the impact of such orientation on economic growth and labour productivity.

Within the framework of the Committee on Agricultural Problems, the FAO/ECE Working Party on Agrarian Structure and Farm Rationalization has been directing its attention to a review of scientific and technical progress in agriculture, its impact on the rational use of natural resources, and the organization of the use of natural resources in agriculture. At the fifth session of the Working Party, to be held in October 1982, economic and social aspects of the rational use of natural resources will be discussed on the basis of case studies prepared by government rapporteurs (ECE/AGRI/57).

Also within the framework of the Committee on Agricultural Problems, the FAO/ECE Working Party on the Mechanization of agriculture is concerned with the environmental aspects of agriculture and its mechanization. A report on the role of agricultural mechanization in preserving and improving the environment under conditions of cropping in flat areas was issued in 1979 and a similar report for hilly and mountainous areas was discussed at the twenty-seventh session of the Working Party in October 1981. Specific questions already taken up include methods and equipment for the application of chemical products in agriculture (report issued in 1980) and equipment and techniques for manure treatment (a general report on this subject was issued in 1977, a report on swine waste handling is being finalized and a report on cleaning, handling and drying of poultry manure was discussed in October 1981). A new topic, discussed for the first time in October 1981, relates to modern tillage techniques and implements. It is expected that the Working Party will continue to review specific environmental aspects of agricultural mechanization in the years to come.

Within the framework of the Timber Committee, the Joint FAO/ECE Working Party on Forest Economics and Statistics is examining methods of quantifying and evaluating the environmental benefits of forests with a view to the assessment of the contribution of forests to society. Methods of calculating compensation for the environmental services of forests and procedures to facilitate comparisons between countries are also being considered (ECE/TIM/17). A special enquiry on the role of forests in supplying environmental and other non-wood goods and services is being circulated to countries as part of a general enquiry on the forest resource base. Preliminary results will be submitted to the 1983 session of the Working Party.

The Chemical Industry Committee is analysing the present state of production and consumption of various types of pesticides and procedures for their use to minimize adverse impacts. A second *ad hoc* meeting, in November 1981, reviewed the first draft of the study (ECE/CHEM/33).

The Coal Committee is dealing with certain environmental problems resulting from industrial activities related to mining, processing, transport and use of coal. Its Group of Experts on Opencast Mines is conducting studies related to increased environmental protection and in particular to land reclamation problems (ECE/COAL/47). In 1983, a Symposium on the Utilization of Waste from Coal Mining and Preparation will be held in Hungary (COAL/GE. 3/R. 41 and Add.1, and COAL/R. 67).

The Committee on Electric Power is examing the technical and economic aspects of environmental problems relating to the production, transmission and distribution of electric power and possible means for resolving these problems. At their tenth session (9 to 11 November 1981), the group of Experts on the Relationship between Electricity and the environment considered the application of an international comparison of standards relating to methods of calculating pollution levels and measuring and monitoring pollution caused by electric power stations. In addition, the problems of land use and land-use planning posed by the establishment of power generating stations and transmission line systems was considered, as well as specific problems of nuclear power and the environment (ECE/EP/35; EP/GE.4/14).

The Committee on Gas has been carrying out consultations on the opportunity of convening a second Symposium on the Gas Industry and Environment, and on the advisability of organizing the symposium in close co-operation with the Committees on Coal and Electric Power and the Senior Adviser to ECE Governments on Environmental Problems (project 07.4.1, ECE/GAS/50).

The Senior Advisers to ECE Governments on Science and Technology will be holding a Seminar on the integrated utilization of low-calorific-value fuels in Moscow (USSR) in early 1983. As part of the programme for this Seminar, it is intended that detailed consideration be given to the environmental implications of the utilization of these fuels. Plans are also under way for the convening of a symposium on rational and integrated utilization of raw materials including biomas and waste recycling. The Senior Advisers to ECE Governments on Environmental Problems will be invited to co-operate in the preparations for this Seminar.

In the secretariat study on "Energy Problems and Co-operation in the ECE Region" (ENERGY/R.20), submitted to the fourth session of the Senior Advisers to ECE Governments on Energy (23 to 27 November 1981), attention was drawn to the environmental implications of emerging energy strategies and developments.

The Conference of European Statisticians, at its last plenary session, considerably modified and expanded its programme of work in the field of environment statistics (see annex I to ECE/CES/18). The comparative survey of national compilations of environment statistics will be continued upon request by interested countries. Studies of the relationship between environment and social (including demographic) statistics, the relationship between frameworks for environment statistics and reports on the state of the envirionment are being prepared. The Conference is attempting to elaborate concepts, definitions and classifications to be used for the study of environmental phenomena. Among particular areas of interest are statistics of land use, water use and quality, and air quality. In addition, the Conference deals with general methodological problems in the development and use of environment statistics (ECE/CES/18).

The Inland Transport Committee is currently involved with developmental studies of the Trans-European North-South Motorway, an ECE project involving the participation of 10 member countries in the development of a network of high capacity roads from the Baltic coast of Poland to south-east Europe, covering a distance of some 10,000 kilometres. The countries concerned have established a framework for the joint study of selected problems, including environmental impact assessment, connected with the construction of the motorway (ECE/TRANS/44).

Within the newly created Working Party on Engineering Industries and Automation, special attention is being given to environmental and resource-saving problems. Recently, a study on Development of Airborne Equipment to Intensify World Food Production has been completed. The study analyses, among other issues, the influence of different types of chemicals, and especially pesticides, on the environment, including the most appropriate techniques for their application. The Working Party's current programme of work includes a study on measures for improving engineering equipment with a view to more effective energy use, centering on the present state and future trends in production and application of engineering equipment and means of automation aimed at saving energy resources. It is also anticipated that a seminar on the topic may be held at a later stage taking into account work done under the auspices of other ECE bodies. Another study is scheduled in the near future on the production of engineering equipment for preventing pollution, particularly that associated with water, and the application of automation for monitoring the environment.

The Committee on Water Problems has entrusted a task force of government rapporteurs with continuing work on the comparison and analysis of methods for defining the various rational standards of water quality, including effluent standards. The Committee is expected to start work at a later stage on methods for assessing the economic impact of water pollution due to industrial and urban activities as well as recreation and amenities (ECE/WATER/25).

The activities of various ECE bodies indicate an increasing awareness that environmental problems are of a comprehensive character and must be taken into account in the processes of planning and decision-making. It is to be noted that ECE offers a particularly advantageous platform to implement impact assessment activities, relying on specific Principal Subsidiary Bodies in their individual generic areas while having the environmental base, within the programme of the Senior Advisers to ECE Governments on Environmental Problems, from which to study and formulate guiding principles on the utility of environmental impact assessment as a public policy management instrument.

SPECIFIC WORK UNDERTAKEN BY THE SENIOR ADVISERS TO ECE GOVERNMENTS ON ENVIRONMENTAL PROBLEMS AND THE COMMITTEE ON HOUSING, BUILDING AND PLANNING

The Committee on Housing, Building and Planning, together with the Senior Advisers to ECE Governments on Environmental Problems, sponsored a Seminar on Integrated Physical, Socio-economic and Environmental Planning in Bergen (Norway) in June 1979. Later that year, the Senior Advisers to ECE Governments on Environmental Problems organized a Seminar on Environmental Impact Assessment in Villach (Austria), and a Seminar on Alternative Patterns of Development and Lifestyles in Ljubljana (Yugoslavia). These three closely related seminars may be characterized as follows:

The Bergen seminar focused on the now generally accepted principle that planning for social development, economic growth and environmental improvement cannot be carried out as separate exercises: integrated planning is essential. Yet it is easier to recognize the need for such integrated planning than to devise policies, institutions and systems ensuring that such integrated planning is being carried out on a continuing basis; it was with this problem that the participants in the Bergen seminar were primarily concerned.

One of the techniques used in integrated planning is environmental impact assessment and, in recent years, this has been made, in many countries, a requirement by Governments and other bodies as an aid to decision-making. Environmental impact takes many forms, however, and the Villach seminar provided an opportunity for ECE countries to share their experience in the use of this technique as well as to examine how it might be improved.

There are signs, however, that integrated planning, using environmental impact assessment and other techniques, may not be sufficient. If recent patterns of development and lifestyles continue, then in many parts of the ECE region, excessive burdens on resources (e.g. energy, water, land) will be seen, and changes may also be forced upon the region by technological developments (e.g. the electronic revolution) or through increasing competitiveness and economic pressure exerted by countries outside the ECE region. The Ljubljana seminar provided an occasion to discuss the types of development patterns and lifestyles that are required to meet social, economic and environmental needs within the constraints likely in the remainder of this century.

The three seminars, in tackling different aspects of integrated planning, brought out common features which, for the purpose of future work, can be regarded as cornerstones. A brief analysis of the discussions and conclusions reveals the following five major ideas:

(a) The present planning framework appears to be too narrow to do justice to the increasing demands to plan for a better quality of life under environmental and resource constraints. The traditional approach, with each activity being planned and implemented more or less independently, has revealed too many deficiencies. Recognition of the complex links between social change, economic growth and resource constraints has led to a demand for a broader and more flexible planning approach. To manage effectively the process of change, integrated planning has been suggested as a suitable approach. Such an approach seems to hold some promise in developing and dealing with alternative development patterns derived from human needs and from recent technological development trends.

(b) An essential characteristic of integrated planning is the transparency of its process. This means that planning has to develop alternatives which will have to be assessed in the decision-making process. Resource limitations and, in particular energy constraints, as well as considerations for the maintenance and improvement of the human environment and increased attention to human relations and self-fulfilment, make the development of alternatives indispensable. This has made the planning task even more difficult and has given impetus to the development of new planning tools. To assess alternatives, existing techniques such as environmental impact assessment are being further developed and broadened to cover other areas, in particular the social impacts. Governments are already experimenting with new and innovative administrative structures but great efforts still need to be made to fully profit from an integrated planning approach.

(c) The increasing awareness of the limitations of natural resources, in particular energy, has added to the already complex planning situation. To cope with this situation, administrative procedures are being augmented by putting decision-making and implementation on a broader basis; the democratic responsibility of the individual, and the general public as a whole, is increasingly being appealed to, in an effort to make them partners in administrative and political decision-making in the search for a better quality of life. Increased public participation has become a necessity and is now implemented in different ways, according to the country's economic and social systems. Even though the effectiveness of citizen participation is highest in decisions close to the individuals concerned, i.e. at the local level, it cannot be neglected at the regional and national levels, e.g. where energy policies are determined. Beyond the traditional participation in planning, more attention is being given to citizen involvement in the implementation and management of plans concerning their immediate environment, e.g. "do-it-yourself" house improvement and co-operative management of housing areas.

(d) As more and more factors are introduced into the planning process, not only by the consideration of different traditional factors but, in particular, through resource constraints and changes in societal values and behaviour, it is obvious that planning for uncertainty is becoming a predominant issue. Uncertainty has also increased through the transitional character of a number of industrial and, in particular, energy extraction and conversion projects, e.g. production of crude oil. To cope with uncertainty, planning needs to be flexible. Alternatives must reflect dynamic changes in industrial location, social behaviour, etc., rather than representing static points in time. Alternatives must also cover the macro as well as the micro planning level to assess the implications of planning policies and to allow them to become effective in site-specific projects which will be carried out despite uncertainty at the macro level.

(e) To make integrated planning a reality, an innovative spirit is needed to co-ordinate forces in a positive and desirable direction. It is now realized that comprehensiveness has its limits and it has been suggested, therefore, that the integrated planning approach should be applied selectively in priority areas (project approach). Innovations will, necessarily, lead to an evaluation of existing administrative structures that will enable a project to be planned, implemented and managed in an integrated manner. This may result in institutional arrangements that are more adapted to the new planning task.

All three seminars have indicated that we have reached the end of the epoch where, in general, sectoral planning and decision-making were sufficient. Limits to economic development, and increasing constraints on non-renewable resources as well as increasing public awareness, make planning more difficult. The principles for a new approach—integrated planning—have been clearly spelled out (flexible approach, development of alternatives, public participation, institutional structure).

PRESENT WORK OF THE SENIOR ADVISERS TO ECE GOVERNMENTS ON ENVIRONMENTAL PROBLEMS

ECE Governments are rapidly coming to the conclusion that the formulation of development policy, based on economic and social goals, must also at the outset and concomitently—include environmental concerns and aspirations; that environmental goals are not to be treated as constraints to be applied in a *post facto* process. The integrated planning and management of modern industrial society depends on the development of a framework which is flexible and adaptive to the pressures and resiliencies of the principal array of public concerns: the economic social and environmental elements. It is within this framework that environmental impact assessment must function.

At their ninth session (10 to 13 February 1981), the Senior Adviser to ECE Governments on Environmental Problems decided to hold an *ad hoc* meeting to define their future activities on environmental impact assessment. The meeting was held from 25th to 27th November 1981 and specifically considered the development of a suggested programme of work. Provided that the proposal for future work has been agreed upon at the tenth session of the Senior Advisers (February 1982), the work on environmental impact assessment will continue.

Concepts, Frameworks and Methodological Problems of Environmental Impact Assessment

The experience of national systems of environmental impact assessment will be examined in the light of the possibility of deriving commonly considered principles and characteristics for an effective, prototypical national process. The economic assessment of environmental damage, social cost/benefit analysis and risk assessment are among elements of the environmental impact assessment system which require further elaboration. General methodological approach and techniques of structuring the assessment process and assessing specific impacts may be studied with the intention of developing recommendations of use at the national level.

Integrated Planning and Management of Development and the Incorporation of Environmental Impact Assessments

The utility of environmental impact assessment requires further exploration and analysis to determine how it may be placed within the framework of integrated planning and management at various phases and levels. Not only does the incorporation problem exist but there is also a necessary restructuring of the impact assessment process to be undertaken. The original industrialized nation concept of a reactive assessment of the effects of pollution, habitat destruction and similar concerns must give way to the notion of an adaptive process for development which includes concern for sustainable resource use and the mitigation of development costs whilst at the same time maintaining the integrity of ecological function. Given developments elsewhere in the programme of the Senior Advisers, as well as that of other ECE Principal Subsidiary Bodies, it is suggested that a productive focus would involve the elaboration of those steps and procedures important in order to assess the environmental aspects of technological development, in particular those technologies intended as environmentally sound, energy- and material-saving processes.

Environmental Impact Assessment of Energy Alernatives

In the pursuit of durable, indigenous sources of energy, new and renewable sources of energy are being regarded by ECE Governments as prime candidates which may, perhaps, minimize environmental and safety problems while providing the long-term security of supply which is so urgent. Work in this area will focus initially on the exploitation and use of the new and renewable sources of energy in co-operation with the work programmes of other ECE bodies where the conventional fuels are already being considered. Activities will be directed towards the assessment of environmental impacts of alternative commercial-scale production and use patterns of selected new and renewable sources of energy, including biomas and solar radiation.

Environmental Impact Assessment as an Instrument for the Resolution of Transboundary Environmental Problems

The possibilities for establishing an international framework providing alternative and interchangeable bases for adapting national systems to the need and requirements of other national systems of the region need to be considered. The notion of a common set of concepts, criteria and approaches will be examined with respect to transboundary pollution problems and shared natural resources. As a first step, an analysis might be carried out on the effectiveness of existing national procedures, to ensure that potential impacts on the environment outside national jurisdiction receive adequate attention. In addition, a study on relevant existing international arrangements would appear to be useful.

PRESENT WORK OF THE COMMITTEE ON HOUSING, BUILDING AND PLANNING

The Seminar on Integrated Physical, Socio-economic and Environmental Planning held in Bergen (Norway) concluded that:

(a) Integrated planning is a process for public policy-making which seeks to improve the quality of life. It implies an understanding of the complexity of modern society and of the major factors and their dynamic relations in a particular situation. It is concerned with conflicts between different goals of society and it seeks to set, through wide participation, parameters within which changes can take place.

(b) Integrated planning seeks:
- to provide an equitable balance between different social interests;
- to secure an efficient use of resources by increasing the rationality of the planning process;
- to achieve socially-acceptable decisions by encouraging public participation;
- to improve efficiency through a more coherent approach to policy making;
- to promote the fuller consideration of alternatives in the planning process by widening the number of factors considered; and
- to take into account uncertainty by providing for flexibility and continuity in the planning process.

(c) The process of integrated planning is a continuous process of related sequential and simultaneous actions. It provides structural framework within which to consider and guide dynamic changes and development, rather than being a prescribed set of routines and procedures. For these reasons, it should possess a substantial management element. It is also a guide to action which, unlike comprehensive planning, concentrates selectively on key issues and, unlike co-ordinated planning, does not seek to impose a rigid framework at any stage of the process.

(d) From the beginning, integrated planning has to be an open process to ensure that the full range of public and individual interests and subsequent changes in their nature and relative priority are properly considered. It can adopt sets of objectives which need not be mutually exclusive and can be pursued at the same time in the same area by different groups of society. To work effectively it must adopt a language which allows those involved in the process to communicate adequately among themselves, with other institutions and with the public.

(e) Integration is required between different levels of organization, between different interests at the same scale and between different groups and priorities. For the successful implementation of integrated planning, wide-spread public participation is essential during this stage because only then can the full impact of change be evaluated. To evaluate the effective implementation of integrated planning, the achievement of objectives contained in the plan must be measurable, although not necessarily in quantitative terms.

(f) To implement integrated planning, organizational structures are needed to provide a basis for continuity. Therefore, the development of integrated planning in an institutional context previously established for sectoral planning usually requires changes in administrative and organizational structures. It is important that funding and implementation agencies be part of integrated planning to ensure that policies and plans are effectively carried through.

Following the discussions and the conclusions reached at the Bergen Seminar, the Committee on Housing, Building and Planning decided in 1980 to carry out the second phase of the project. It was considered desirable that a project should be undertaken in which experience could be exchanged on practical problems encountered in the planning and implementation of large-scale infrastructural and industrial projects. Such a discussion would be most fruitful if centred on site-specific projects which have an immediate effect on individuals, thus eliciting essential public participation, but which, at the same time, allow for the analysis of the planning and implementation process and the corresponding administrative structure. The project should furthermore try to tackle the problems of uncertainty which result, for example, from the exploitation of limited natural resources or the development and redevelopment of industrial areas and major technical infrastructure. In this way the project will deal with environmental concerns, the need to develop local and regional economies, its impact on the employment situation and, finally, the effect it would have on the improvement of the quality of life in human settlements.

As a starting point, the delegations of Poland and the United Kingdom have prepared case studies. The Polish delegation has described the planning of the Vistula River Development Programme (HBP/SEM.31/COM.1) and the United Kingdom delegation has illustrated the planning and the start of the implementation phase of the St. Fergus Gas Terminal at Peterhead, Scotland (HBP/SEM.31/COM.2). On the basis of these two case studies and assisted by a set of guidelines, national delegations have been invited to respond with their own case studies in the two major areas of infrastructural and industrial development projects. The lessons learned from site-specific projects will be discussed at a seminar to be held in Spring 1983. The seminar will also take stock of national policy developments which have occurred since the Bergen Seminar in 1979. It is anticipated that the seminar will formulate a number of conclusions emerging from the site-specific case studies.

CHAPTER 2

Environmental Management: Issues of an Integrated Planning Approach

F. G. Müller

INTRODUCTION

DURING the last decade, the debate of issues relating to environmental management has been shifting its focus. Instead of arguing and promoting the importance of particular environmental policy choices, the debate now centres around environmental issues in relation to the problem of priorities for society. In particular, the aim is to determine how much environmental protection society can afford and implement, consistent with interdependent goals for other competing, or even conflicting, social concerns, such as economic development, price stability, equitable income distribution and energy availability. In this reoriented debate, the current environmental evaluation and management practices are only of limited relevance. They are structured according to the traditional scientific approach to society's problems, namely, the classical notion of separable problems and, as a consequence, they fail to offer an integrated environmental policy within a socio-economic context.

Although it is recognized that decisions about economic development should be based on environmental considerations as well as economic and social benefits and costs, there is no consensus on how best to incorporate ecological and environmental values into the decision-making process (UN, 1972). There is general agreement that their incorporation is not usually executed very successfully. To fill this gap, various specialized agencies of the United Nations system have been involved in designing comprehensive planning schemes.[1] However, the methodology of integrated physical, socio-economic and environmental planning is still a relatively new and evolving concept of economic, environmental and social management. The purpose of this chapter is to illustrate some of the issues of the integrated planning approach since, at the operational level, it still raises many difficult methodological problems which require further research.

[1]Some of the UN documentation on this subject can be found listed in the References section at the end of this chapter.

ECOLOGICAL AND SOCIO-ECONOMIC SYSTEMS

Ecology is a biological science dealing with the relationship between organisms and their surroundings. The main objective of environmental management is to minimize the impact of man's activities on the physical and biological environment through a perspective of the ecosystem (Odum, 1981).

The significance of a specific ecosystem is a subjective, value judgement of its biological importance. To many ecologists, biological importance means closeness to a pristine and/or virgin state. In a broader perpective, however, ecological significance is generally considered to be a function of the importance of a particular ecological unit to man's activity. These two concepts may be reconciled, by intially defining ecological importance on the basis of closeness to pristine conditions and then modifying that definition in specific cases to take account of explicitly stated societal decisions or preferences for different ecological states (Cooper and Bedler, 1980).[2]

The natural ecosystem, however, has no intervening 'markets' (or central planning) as such in its circulatory process of energy flows, food chains and the like, mainly because they are self-regulating, the reason being that the first and second laws of thermodynamics establish an equilibrium in energy conversion, nutrient cycles and species population. In contrast, human needs as a whole have no such constraints, rather they appear to be almost infinite, and thus society has to allocate the scarce resources through markets in the circular flows of human production and consumption to substitute for the regulatory functions of natural and/or physical laws.

This ecosystem is basically a closed system, except for incoming solar energy. Consequently, the nutrient-species processes are self-contained and circulatory within themselves. The socio-economic system is, however, an open system with newly created wants and desires appearing on the demand side of the market, resulting in further demands upon the resource base and, finally, generating increasing quantities and qualities of wastes disposed into the ecosystem.[3] The socio-economic system is based solidly on the material base which is subject to finite constraints. Due to these resource constraints, the economic process has a unidirectional, irrevocable evolution. In the economic system, only money circulates within and between economic sectors (Georgescu-Roegen, 1976. p. 56). At present, a 'free' exchange system exists between the ecosystem and the socio-ecosystem. This 'free' exchange, coupled with the apparent infinity of human wants, is what has caused the ecological crisis. A result of this 'free' exchange system has been the failure of the

[2]Some aspects of ecological importance to man can be measured, such as the potential yield of harvestable timber; other aspects, like aesthetic values, are not, or hardly, commensurable.

[3]The open economic system has no regulating function in its relation to the ecosystem, either in the form of markets or natural laws, unless poisoning mankind into extinction. Since the natural laws of toxicity (poisoning mankind) are not acceptable, society needs some other forms of control.

established market system (and planning system) and a misallocation of resources, causing an over-production of wastes and, in the last analysis, a threat to human survival. Consequently, the need for a new approach to decision-making is required—an integrated, environmental management approach, profoundly interdisciplinary and capable of responding to society's assessment of environmental quality and anxiety over ecological damage.

ENVIRONMENTAL MANAGEMENT: AN INTEGRATED APPROACH

Environmental management based on an integrated planning approach consists of managing human activities in order to achieve an acceptable (depending upon the values of the society) balance between the quality of the human environment and the quality of the natural environment. This complex management task requires the environmental manager to be aware of, and to recognize, the key factors which constrain his activities. These factors can be identified as social, technical, political, legal, ecological and economic. An understanding of these interrelation-ships requires multidisciplinary knowledge and an interdisciplinary, integrated planning approach to develop public policy programmes that enhance and achieve the politically desired quality of life.

In order to facilitate the comprehension of environmental management as a complex and interrelated system, it is necessary to identify the main subsystems involved, i.e. those subsystems whose relationships and interdependencies define and dominate the overall system. This is a complex task which can be accomplished at various levels of abstraction and detail. The intention here is to provide a framework only for discussion, not a highly detailed analysis or major treatise in the field of environmental management. As a point of departure, an integrated planning approach for environmental management must consider the dominant factors of the overall system. Each factor within the system will be analyzed for its interaction with the decision-making function, with the focus on illustrating the issues involved in designing an integrated planning approach.

Information Problems of Large-scale Planning Models

The increasing interlinkages between ecological and socio-economic sectors of society have placed a variety of pressures upon environmental decision-makers concerned in all levels of government. The view of environmental management as an interrelationship of complex subsystems presents a difficult role for the decision-maker and planner, namely, management is facing a large range of uncertainties and ambiguities, usually not considered under a more restricted economic decision model (Petak, 1980). In this context, management must attempt to reduce uncertainty, while trying to maintain the flexibility necessary to respond to changing social and political values and preferences, and, in addition, to integrate the activities and balance the demands of these various subsystems. The integrated planning and decision-making process involves choosing among alternatives and is a crucial factor in managerial performance, system effectiveness, and ultimate environmental quality.

Many of the present environmental difficulties are the direct result of alternative decision selections derived from too narrow decision criteria, ignoring the impacts on the ecosystem and the final social costs. As a consequence of this, at least three trends can be observed in different countries where environmental planners were obliged to re-evaluate their roles:

(a) The emerging emphasis on public participation in the planning and evaluation process has, in the last decade, moved the decision-making process from professional dominance to the point where persons from all levels of society can take part. Citizens have given up their passive role and are now claiming full participation in environmental planning and decision-making. Consequently, the professional manager has to learn to adapt to and reconcile divergent interests and opinions.

(b) National environmental acts, which require an impact-assessment component to be included in all medium and large-scale projects, have caused the planner to get involved in comprehensive environmental evaluation and in understanding the expected socio-environmental impacts prior to accepting a project.

(c) A change from focusing on new designs for the environment to a broader interest in prevailing environments, their conservation, enhancement and management, is taking place. Environmental management, which pays more attention to non-degradation and monitoring of the environment over time, is viewed as an alternative to environmental improvements generated by single-purpose, one-time interventions.

Responsible decision-making and planning has the arduous task of identifying the many alternatives and trying to predict what consequences will follow upon the choice of each alternative. The prospect of constructing such a model of alternative choices is formidable because many of the ecological and physical interactions in nature are unknown. As a result, the information available for the integrated planning model is only a fraction of what would be required to illustrate the real environment. At best, some inferences can be drawn from such a comprehensive model and submitted to the test of debate from intuition, experience and common belief. In other words, a large-scale integrated decision model should enlarge the range of alternatives. However, as soon as one attempts to set up such a large-scale model, one is confronted with what some authors call the "ethics of the whole system", namely the problem that models do not mean anything unless they contain reliable information (Churchman, 1968). But one is unable to determine what information is reliable unless one possesses the knowledge of how all components interrelate to the whole system. The dilemma is that one requires correct information initially to construct a model, but one also requires the model first in order to ask the appropriate questions and to get the relevant information.

Thus, the dilemma is how to break out of this vicious circle. From past experience, one guide is to note that present environmental problems appear to result from basing decisions on a narrow incrementalism, focusing on short-term utility maximization. A pragmatic guideline is to design as large a conceptual framework of man's relation to nature as we know how and then to submit it to the test of data, experience and controversy. Thus, in order for the decision-maker to function in a manner relevant to policy (namely, to be both project/programme assessor and advocate his position), the integrated planning process must effectively treat input from all subsystems on perceptions of environmental amenities and pollution, ecological criteria and standards, materials balance assessments, environmental and socio-economic indicators, environmental monitoring results, cost/risk/benefit assessment and institutional/behavioral changes. This integrated approach will require each decision-making subsystem to develop and maintain comprehensive data bases. Ideally, these data should flow from each sub-system, then be compiled in terms of indices that allow the aggregation of the relevant data, be stored as part of the decision-maker's information system and form the base of the integrated planning process.

Societal Subsystem

The responsibility of the environmental decision-maker and planner is to serve the major goals outlined in each country's national environmental policy act. The act usually declares a policy to support and promote general welfare and to create and maintain conditions under which man's activities and nature can co-exist in harmony. This policy statement means different things to different people and no consensus exists as to what is meant in objective terms. Thus, understanding the varying attitudes, values, and perceptions of those in the societal subsystem is critical if the integrated planning process is to be effective. However, before quantitative techniques can be applied, it is necessary to specify the goals in a quantitative form. Therefore, goal-definition effectively becomes a part of the decision-making and planning process.

One of the first problems is to decide whose goals should be considered. This is relatively easy to determine in a dictatorship or one-party state, although the goals of the elite cannot totally ignore the needs of the people, i.e.

to keep the people 'content' or at least under control. In a democracy, planning goals should reflect the goals of society at large and there are at least three possible ways of determining these: referendum, sample survey and through elected representatives. Each of these ways has its particular characteristics but it is conceivable that society could reach a compromise to accept one of them as the standard, public choice mechanism (Mueller, 1979). However, this does not yet eliminate the difficulty as it is still necessary to determine the geographical, local and/or spatial distribution for the output of the proposed project. For example, in an anti-noise programme, should the whole country be considered, or one or two regions, or only those directly affected? These are decisions of an essentially political nature which vary in complexity and features from project to project. Again, it may be conceivable that guidelines could emerge to handle this difficulty but it seems unlikely. Political pressure groups do not tend to give up their influence over a politically made decision voluntarily in order to make their actions more predictable. Assuming it is possible to select a group whose goals will form the social goals, two basic problems remain; firstly, how are the goals of each individual to be determined and secondly, how are individual goals to be aggregated to form social goals?

The goals of an individual depend on the context; a person may, for example, favour the preservation of wilderness as against recreational development. Furthermore, goals change with time, knowledge and circumstances. At present, due to the prevailing economic recession, environmental conservation ranks relatively low in comparison to previous years. Thus, to obtain a valid set of goals for any individual, it would be necessary to ensure that he was fully informed about the project/programme and its alternatives. However, in reality, no person has complete information and perfect foresight. This requirement could, therefore, restrict the representativeness of the goal-setting group by limiting it to the 'informed and/or educated' elite.

The 'free-rider' syndrome also questions the validity of goals revealed by individuals who may use strategic answers to improve their benefits (Haefele, 1973). For example, an individual may rank wildlife conservation as a low priority goal because he assumes it will be provided anyway and he fears he might have to share the cost from this programme if he reveals his true preferences. From these considerations it follows that the goals would have to be determined for every project/programme. The context-dependence of the goals implies that it would not be possible to change the range of alternatives considered without re-establishing the goals. Thus, the goals would have to be established towards the end of planning the project rather than at the beginning and, consequently, would not be available to guide the environmental manager and integrated planner.

In his Impossibility Theorem, Arrow has shown that under a set of simple and apparently innocuous assumptions, it is logically impossible to obtain a consistent social preference order derived from the set of individual preferences if there are more than two alternatives (Arrow, 1963). Apparently, up to now, no satisfactory method has been developed to solve this problem and further research is required in this area. If the assumptions of Arrow's Theorem are accepted, it follows that any set of social goals derived from individual goals will be invalid, regardless of the technique and/or method of aggregation chosen. Thus, the concept of a definable set of generally accepted social goals becomes controversial.

Even if a set of social goals can be identified and ranked, there is still the problem of quantifying all goals in commensurate units to enable a usable objective function to be derived. Goals and objectives are generally not directly measurable and criteria have to be selected that are measurable and, in some sense, reflect the goals being sought. Great care is required in the choice of these criteria in order not to bias the objective function towards one type of solution.[4] So far, this discussion suggests that there are several problems in the methods of defining social goals. In a democracy, there are interest groups in the community with widely divergent objectives and values. Such groups will always prefer direct influence at the decision-making and planning level to less tangible influence over the definition of social goals, i.e. value judgements are inherent in social goal-setting and, apparently, cannot be separated from the decision-making and planning process itself.

Criteria for Trade-off Analysis

The process of planning and decision-making may be viewed as a balancing of probable gains aganst losses. The theory of choices concerns the means by which human decision-making can obtain a relative gain minus the least relative loss. This decision function may be described as trade-off analysis—the subjective calculation by which the probable gains minus the probable losses yield a preference. If incorrect information causes the decision to yield larger relative losses (i.e. opportunity costs) than gains, the human organism moves from its present state toward one of greater entropy. This implies that the decision was not optimal because it accumulates more ecological damage to the organism than energy; in other words, the decision lost time in a life-span sense. To gain or lose time is the essence of a trade-off decision for human organisms, for time is life itself. Pollution in the environment results in ecological damage which reduces life-span; this is a loss of time or 'lifetime'. Furthermore, time is interchangeable with energy because it is the storage of potential energy by the organism which provides its lifetime potential.

This concept can also be applied to economics. As economists have pointed out, exchange should not be viewed as a means of obtaining products,

[4]For example, the goal of a cost-efficient mass transit system could be evaluated by cost/passenger km or cost per ride. The first ratio favours long, high-capacity hauls and the second is biased in favour of a dense transportation network. Similarly, governments have been able to 'solve' social problems by changing the criteria by which they were determined.

but rather of obtaining the services of products as the objects of exchange: for example, a consumer does not buy food *per se,* but the services of food; nor a dwelling, rather the services of a dwelling. The services of these products are specific forms of energy; the services of food are in caloric yield to maintain a steady state in the human organism, while the services of a dwelling are heat-protection to conserve the potential energy of the body.

The energy potential made available to its user is the service of all material products. It is this very acquisition of energy utilization which is central to economic activities and exchange (Georgescu-Roegen, 1976. p.8). The utility of an object is expressed in energy potential and, consequently, energy services are a source of human satisfaction because the organism then avoids using up its own energy (low entropy); by means of preserving energy reserves, an organism is able to extend its time or its life. Thus, the common denominator in ecological trade-off analysis is time, while in economic trade-off calculations, the net gain (or profit) is usually expressed in monetary units. The ecosystem does not deal in monetary units; instead, its media are energy, matter and time, where time is the variable rate of conversions between energy and matter. For example, if environmental health is taken as the subject of ecological studies, then the criteria for trade-off analysis are the species population affected and their life-spans, i.e. the time dimension.[5]

The economic planning and decision process is not constructed to recognize these trade-offs because there are no ecological parameters in the decision mechanism. At present, this mechanism recognizes only material and energy inputs being converted into services to satisfy human wants, with a net profit for the conversion service production. Consequently, this decision mechanism favours providing nutrients and energy reserves to the human species at the expense of other species. This, however, implies that the disadvantaged species succumb, as they are described energy reserves, and this reduces the remaining available food supply for human survival. In this light, the concept of profit from economic activities becomes a fictitious gain, generated by ignoring the ecological costs. If all costs and ecological damage are fully accounted for, net profit as a criterion for decision-making is a logical impossibility because it is contrary to the second law of thermodynamics.[6] In effect, the concept of net profit shifts the ecological damage to other organisms who then bear the compensatory losses by a deficiency in energy and time; as Wiener stated: "It may be a long time yet before our civilization and our human race perish, though perish they will." (Wiener, 1950). The objective of ecological trade-off analysis, as part of the integrated planning approach, is to determine which species pay in energy, time and loss of life for the gains sought in economic satisfaction by humans. This is basically an entropy accounting approach.

Despite the validity of the entropy law in evolutionary terms, it does not

[5]This implies that one species' gain in nutrients or access to energy reserves is another species' loss.

[6]Energy may be transformed from a lower to a higher state of entropy, but always at a loss.

reveal the intensity of this process over time; therefore, it is undermined and, consequently, does not offer a reliable guideline for medium- and long-term planning problems. Even though the trade-off analysis is a means of trying to make decision-makers and planners aware of the entropic consequences of their actions, it seems that traditional tools, despite their shortcomings, have not yet outlived their usefulness.

Quantitative Methods

Environmental Impact Assessment

There are a number of quantitative methods of assessment available to decision-makers and planners. To accomplish its full purpose, an assessment must (1) identify, (2) measure and (3) evaluate environmental impacts. Usually only the first two steps are accomplished to any degree in the majority of assessment studies. The emphasis is placed on an enumeration of possible impacts and on data collection and display, with discussion usually centred on how well impacts are recorded and measured. Far less provision is made for the weighing of relative values, which should be the main purpose of this assessment activity (Bisset, 1978). The demand for environmental impact studies (EIS) has generated proposals for so-called methodologies to satisfy the planning or decision-makers' requirements. Most of these proposals suggest or prescribe a list of data requirements and most differ only in their suggestions for presentation of the information (Munn, 1976). Basically, they can be arranged into four groups: (a) checklists, (b) matrices, (c) networks and (d) map overlays. Occasionally, computer simulations are mentioned as another alternative. The checklist is a common feature and usually represents a catalogue of possible impacts that might be expected from various types of activities. As such, this type of catalogue provides only meagre information about the extent of impact and its relative importance. (Edmunds and Letey; Rowe *et al*, 1978). Matrices instead combine a list of impact possibilities with various project activities that might be associated with or produce such consequences. The purpose of these matrices is more to be explicit in discerning which actions have an impact on which environmental characteristics. Network proposals attempt to describe more comprehensively case and effect relationships, listing impacts and indicating how these are generated through effects on resources. Finally, map overlays try to identify areas of lesser conflict among resource uses and environmentally important values. This is accomplished by superimposing maps describing various uses for values (such as vegetation, water courses or unique wildlife habitat). The method is employed in locational screening of alternative project sites. However, its chief shortcomings are that it fails to differentiate the weight or relative importance between the respective resource uses, nor does it indicate the interrelationships among them. Environmental impact assessment may give assurance to planners that their decisions on programmes and/or projects are made on a broader scope of values than would otherwise be the case. However, the separately conducted study or listing of environmental factors has not been a particularly efficient or convincing approach to including environmental considerations in prevailing development appraisal and planning methods. Implementation of such considerations has frequently been costly and time-consuming. Most of the time, the centre of attention in devising environmental impact requirements has been on the truly massive or unusual project. For such projects, unique environmental procedures may very well be advisable but they become questionable when applied to the numerous routine projects planned annually in nearly every country. Obviously, EISs must be efficient, manageable and operational. Massive and comprehensive EISs appear to be redundant in every instance. Instead, the process should elicit appropriate information and the screening must be responsive to the relative values.

In most cases, EISs were conducted following the directive which centred on data to be generated and the form for presenting this gathered information, avoiding almost entirely any comparison and accommodation of economic and environmental impacts. Sometimes, it was even specified that economic values were not to be included in the EIS. With the emphasis on producing data on purely environmental matters, it is not surprising that the normal response of programme proponents is towards side study, usually contracted-out to people independent of those involved in planning the project itself and with the sole purpose of collecting environmental information. With such a narrow scope, these studies obviously have little focus, less point and do not improve interdisciplinary planning methods.[7]

All methods of assessing the environment are fraught with major difficulties. Chief among these is the assignment of values to the relative significance of predicted changes (Bisset, 1978). Whereas the more traditional cost-benefit analysis (CBA) of development projects lends itself to some measurement in monetary terms, the impact of projects on the ecosystem have not yet been usefully and reliably quantified, in particular on commensurate scales. Deciding what is of greater or lesser importance is a generic problem. Current EIS methods, almost by definition, vary with the evaluator involved and have to be viewed as a personal value judgement. Comparability and reproducibility of inter-project ratings—sometimes even among the specialists them-selves—remain a constant source of concern. Guidelines repeat the need for more objectivity. Furthermore, attempts to aggregate and/or summarize different impacts in some sort of quality index are also inherently arbitrary and highly subjective. An additonal problem with current EIS methods is how to deal effectively with basic unknowns and uncertainties of possible impacts.

[7]It seems that the necessity to show environmental concern to the court in several countries, this perverse tendency to assemble masses of information over relevant information was even reinforced, since the environmental interest lawyer will not challenge the adequacy of the EIS if its "weight" is overwhelming enough.

The present understanding of causal factors in the environment is far from complete.

The Cost-benefit Algorithm

Cost-benefit analysis is designed to make possible informed value judgements to assist the decision-makers and planners in ranking investment projects according to their economic efficiency. Planners require both qualitative and quantitative considerations, but no model or decision rule can generate the 'best' decision by itself. Economists hope that cost-benefit analysis can guide the final decision by highlighting the uncertainties and classifying the assumptions and calculations involved. (Hollick, 1981; Nash *et al,* 1975). Although specific formulation differs from case to case, cost-benefit analysis usually involves some combination of the following steps:

(a) specify the goals and objectives of the project(s);
(b) identify the expected benefits and costs for each alternative;
(c) define the unit of measure applicable to each of the outputs and estimate the quantitative number of units;
(d) apply monetary terms as common index of relative value to allow comparability and additivity;
(e) introduce weighting factors in recognition that value judgement will be part of analysis. Weights are employed to adjust the estimates of monetary value for non-market considerations of social, cultural and ecological aspects of the evaluation process;
(f) estimate, aggregate and discount benefits and costs of the identified consequences of the project; and
(g) employ ranking criteria to evaluate alternatives.

Traditionally, cost-benefit analysis has proved to be quite valuable to resource managers in general and to environmental managers in particular. However, it has to be noted that cost-benefit analysis is not a prescriptive tool. It is most useful, and less controversial, for the application of relatively small, single-purpose projects with few externa-lities. Evidently, environmental managers rarely face such situations. Substantial shortcomings are associated with the cost-benefit analysis such as the relative ease with which the expected flow of benefits and costs can be manipulated according to the planner's interest; the necessity of restricting the scope of the project since large projects and programmes tend to alter the entire range of relative values and outputs; selection of an appropriate rate of social discount; determination of the proper estimates for values to be considered in the case of imperfect or non-existent market evaluations (such as environmental quality); justification for the inclusion or exclusion of certain benefits and costs; and the assessment of income redistribution effects.

Some of these shortcomings (not all and only to a certain degree) can be adjusted reasonably well. However, regardless of the degree of sophistication of the cost-benefit analysis approach applied, no present cost-benefit analysis is equipped to eliminate the need for substantive value judgements in the final stage of the planning and decision-making process. Instead, the planners have to rely on interdisciplinary, orientated teams to assess the value issues which are inadequately reflected in the cost-benefit framework.

Strategy for the Use of Quantitative Methods

While EIS methods are still developing and procedures have to be considered tentative, it now appears obvious that efforts and research should not simply be directed towards producing more ingenious, sophisticated and longer lists of types of environmental impacts that may result from a project. A checklist by itself only provides the enumerations of impacts. It may be a necessary first step; but it does not mean that such enumerations are explicitly considered in the evaluation of a project's desirability. In fact, the EIS procedures currently in use are not methodologies of EIS at all, but rather quite undirected, basically an inventory of physical and ecological information with guidelines for detailing changes. Despite procedural shortcomings of EIS studies, they have occasionally been important in some decision to alter and/or prevent the implementation of projects.

This separation of EIS from economic evaluation prevents environmental impacts being considered at an earlier stage in the planning of the project. It is possible that this separation may tend to make planning decisions sequential by first preparing an economic feasibility study and then, at a later stage, considering the environmental constraints. If this describes the usual planning process, then there is hardly any (or no explicit) trade-off among alternative choices made in the formulation stage of the programme and/or project, thus making substantial inflexibility. Consequently, the burden of proof is such that an extremely convincing case has to be made to cause changes in a project since at this planning stage, the set of alternatives is already reduced and such changes will probably also be more costly. Thus, at the moment, EIS are often viewed more as adversary reports than as instruments to enhance the planning process. The practice of environmental impact accounting, with little or no requirement to consider environmental economic trade-offs, can also lead to inconsistent treatment of environmental impacts across different projects and, therefore, result in wasteful allocations of resources. In fact, without priorities, the request for environmental data collection is very much open-ended, rather expensive and often proves of little benefit to the integrated planning process (Knetsch and Freeman, 1979).

The separation of EIS from project/programme planning may also generate conflict between the development planning team and the EIS team and thus, prove to be rather obstructive to an integrated planning process. What seems to be required for an integrated planning process, if the EIS are to be substantive and useful, are procedures which emphasize explicitly the trade-off between measurable economic gains and more intangible environmental impacts. This connection of environmental effects to economic gains and losses could be achieved by linking the EIS to the criteria of the cost-benefit analysis (CBA) and requiring the necessity of a supported judgement.

Despite the previously mentioned shortcomings, the CBA has one major virtue—an explicit conclusion concerning the expected benefit/cost ratio. As a consequence, the awareness of this feasibility test, and the obligation to demonstrate that the evidence supports the conclusion, influences planning, from the very beginning of the project/programme, to carefully select options that reduce the costs and increase the gains.

The strategy of cost-benefit analysis can also be usefully employed in the assessment of environmental impacts, even though amenities and other uses of environmental resources are difficult and/or cannot be expressed in monetary terms (Weisbrod, 1964; Cicchetti and Freeman, 1971).[8] A positive recommendation for implementing a project implies that any negative environmental impacts are outweighed by beneficial gains to other objectives and considerations, or that any positive environmental impacts outweigh any negative effects on economic welfare or other social-political objectives. Thus, a decision rule of such an integrated planning approach would require that recommendations for decisions to proceed with a project/programme include an explanatory statement that the net economic benefits of the project under consideration exceed any negative environmental impacts, or that the negative economic benefits are exceeded by the expected, positive environmental benefits. In this exercise of value, comparison of the EIS would become central. The integration of these two concepts (EIS and cost-benefit analysis), according to the suggested comparison, would force the decision-makers and planners to demonstrate that the benefits exceed the costs. Furthermore, it also provides a point of orientation for the collection of required information and leads to interdisciplinary co-operation.

Any statement of the relative values would be required to be substantiated by details of evidence and assumptions used to justify proceeding with the project. This evidence should be arranged in such a way as to clearly illustrate a comparison of the net economic gain with any negative environmental effects. This comparison would require the following steps:

(a) to identify the potential effects on the environment associated with the project alternative under consideration;
(b) to assess the nature of values associated with the identified effects and separate quantifiable values (usually expressed in monetary terms) from non-quantifiable values; and
(c) to state necessary assumptions to support a judgement of economic and environmental justification of project choice over others, including the assumptions of how values are expected to change over time in response to changes in demands for both project outputs and uses of natural environment.

An important element in this comparison is the assumption of values concerning the time paths of expected benefits of preserving resources and

[8]However, recent contributions have extended the conventional CBA through the inclusion of the option value concept for preserving natural environments.

alternative development projects. Thus, the analysis would illustrate any expected change in the field of benefits over time from each of the alternative options and show how these assumed growth-paths lead to the conclusion reached about the relative values of environmental and developmental benefits. The respective time path has to include:

(a) the expected life of any facilities related to alternative development;
(b) changes in the demands for any of the commodities of each alternative over time (including any other possible consideration that would affect the flow of benefits over time);
(c) the degree of available substitutes for commodities to be produced as a result of the suggested project (including those stemming from non-development); and finally,
(d) the consequences of any irreversible characteristic of change in the use of the resources being considered (Krutilla and Fisher, 1975).[9]

The necessity of explicit statements on the economic desirability of the project, and the associated judgement that any identified environmental costs are less than the net economic gains (thus presuming it to be the most efficient choice), require the decision-makers and planners to arrange the information and to make an explicit judgement about the values involved. The suggested strategy, of explicitly noting the economic and environmental trade-offs, lends itself to efforts to account for changing values associated with natural areas, or other unique environmental assets. Thus, it seems to be a useful concept for an integrated planning approach of ecological and economic criteria.

Conclusions

While it has been intended to illustrate how an integrated planning approach can be used to conceptualize and model the complex environmental management system, it is also necessary to recognize the existence of several subsystems. Through this integrated planning approach, each of the individual subsystems is able to adapt to impacts and opportunities, as well as to learn from its behaviours. A complete analysis of the environmental management system requires a thorough examination of the interdependencies and interactions of all system elements, the sensitivity of changes to the elements and the process needed to respond to the larger environment. It is evident that the fields of government, economics, social responsibility and ecology are so complex that it is extremely difficult to develop a framework that gives full consideration to all aspects. This discussion has attempted, however, to show a number of problems with quantified decision-making techniques when applied to social decisions, such as are involved in EIS and CBA.

It is important to re-emphasize that such techniques can never be truly objective and free of bias and subjectivity. No amount of procedural reform

[9] This approach was developed and successfully employed by Krutilla and Fisher in the Hells-Canyon Case, USA.

will guarantee a fully adequate and appropriate evaluation and weighting of non-quantified amenity and preservation values. However, shifting the focus of environmental work to a more inclusive project-feasibility test, rather than a separate EIS, should encourage the employment of individuals working on environmental matters throughout project formulation and design stages. The inclusion of environmental effects can also lead to a wider search for alternatives likely to be less costly in social terms and, thus, a more explicit assurance of ecological sustainable development is implied.

The procedure suggested could have a major beneficial impact on the whole integrated planning process of project location, design and justification. Project proponents would be encouraged to develop alternative means and locations, where negative environmental damages are minimized.

For projects having great measurable benefits and few identifiable environmental problems, it will not be difficult to support the assertion that the benefits outweigh the environmental losses and, consequently, this statement will prevent undue delays in implementing the project. For alternatives of more modest economic gain, relative to environmental problems, the case for proceeding with the project will appear more crucial and questionable. Such a discrimination is the aim of the proposed procedure. For projects that have no alternatives (such as, only one feasible dam site) but whose economic gains are also not substantial, relative to environmental damages, the procedure would force the search for very different solutions to fulfil the existing economic and social goals.

One truism that has emerged from this discussion is that the real-world situations, which integrated planning approaches attempt to model, are complex. Up to the present, efforts to control or influence environmental systems have been made with reference to goals which were deeply held only by small, articulate and influential groups of society, in order to generate incremental changes in certain aspects of system operation. Because environmental systems are complex, the results of such planning techniques and control actions often include those which could be broadly 'detrimental' to society as a whole, such as a uniformly grey police state or the effective exclusion of large sections of the population from the most scenic parts of the countryside. There is no easy or quick answer to this ultimate planning and control problem of environmental quality, except to exhort an increasing realism through an integrated planning approach. In this context, three matters are of particular concern: firstly, the goals of the planning process, control and intervention; secondly, the results of such activities; and thirdly, to what degree the goals of planning, control and the intensity and hierarchical level of planning control must be moderated in order to result in broadly beneficial effects. On the one hand, it may appear that it is merely the inadequencies of current systems, conceptions and planning techniques which separate a Kafka-like hell from a Marxist heaven. And on the other hand, it is much more plausible and probable that all but the most broadly humane and delicate of planning approaches and controls over environmental systems may be disastrous. (Bennett and Chorley, 1978). However, ultimately, one must realize that regardless of the planning approach chosen, all decisions have at least to be sanctioned by the holders of political power. No analytical method can ever force the powerful elite to be rational in a broader sense, rather than in a limited political sense, if they do not intend to be. The role of the analyst, manager and planner and their tools is one of assisting those in power to be rational it they so wish.

References

Arrow, K. (1963). *Social Choices and Individual Values,* (2nd ed.). John Wiley & Sons, Inc., New York.

Bennett, R. and Chorley, R. (1978). *Environmental Systems: Philosophy, Analysis and Control.* Methuen, London, p. 533.

Bisset, R. (1978). Quantification, Decision-making and Environmental Impact Assessment in the United Kingdom. *Journal of Environment,* Vol. 7, pp. 52-53.

Churchman, C.W. (1968). *Challenge to Reason.* MacGraw-Hill, New York, p. 161.

Cicchetti, C. and Freeman, F. (1971). Option Demand and Consumer Surplus; Further Comments. *Q.J. of Econ., Vol. 85, 3.*

Cooper, C. and Bedler, P. (1980. Ecological Assessment for Regional Development. *Journal of Environmental Management,* Vol. 10, p. 2.

Edmunds, S. and Letey, J. (1973). *Environmental Administration.* McGraw-Hill, New York, p. 169.

Georgescu-Roegen, N. (1976). *The Entropy Law and Economic Problem in Energy and Economic Myths: Institutional and Analytical Economic Analysis.* Pergamon Press, New York, p. 8, 56.

Haefele, E. (1973). *Representative Government and Environmental Management.* The Johns Hopkins University Press, Baltimore.

Hollick, M. (1981). The Role of Quantitative Decision-making Methods in Environmental Impact Assessment. *Journal of Environmental Management,* Vol. 12, p. 75.

Knetsch, J.L. and Freeman, P.H. (1979). Environmental Economic Assessment in Development Project Planning. *Journal of Environmental Management,* Vol. 9, p. 241.

Krutilla, J. and Fisher, A. (1975). *The Economics of Natural Environments.* The Johns Hopkins University Press, Baltimore, p. 84.

Mueller, D. (1979). *Public Choice.* Cambridge University Press, New York.

Munn, R., Editor. (1976). *Environmental Impact Assessment: Principles and Procedures.* Scope Report 5, Toronto.

Nash, C., Pearce, D. and Stanley, J. (1975). Criteria for Evaluating Project Evaluation Techniques. *Journal of American Institute of Planners,* Vol. 2, p. 86.

Odum, E. (1981). *Fundamentals of Ecology.* W. B. Saunders Co., Philadelphia.

Petak, W. (1980). *Environmental Management: A System Approach.* Vol. 5, No. 5, pp. 213-224.

Rees, W. (1976). *Scope and General Methodologies.* Proceedings of Environmental Impact Assessment Conference, University of British Columbia, Vancouver.

Rowe, P.G., Nixon J., Smith, B.A., Blackburn J., Callaway, G.L. and Gevirtz, J.L. (1978). *Principles for Local Environmental Management,* Ballinger Publishing Co., Cambridge, Massachusetts, p. 41.

UN (1970). *Report of Interregional Seminar on Physical Planning for Tourism Development.* Dubrovnik, Yugoslavia, 19 Oct - 3 Nov 1970. UN ST/TAO/ Ser. C/131.

UN (1972). *Report of the United Nations Conference on the Human Environment.* Stockholm, 5 - 16 June 1972, p. 4.

UN (1976). *Planning and Development of the Tourist Industry in the ECE Region.* United Nations, New York, 1976.

UN (1979). *Seminar on Integrated Physical, Socio-economic and Environmental Planning.* Bergen, Norway, 18 - 22 June 1979. ECE, HEP/Sem. 23/2; ENV/Sem. 12/2.

Weisbrod, B. (1964). Collective Consumption Services of Individual Consumption Goods. *Q.J. Econ.,* Vol. 78, 3.

Weiner, N. (1950). *The Human Use of Human Beings.* (1968 Edition). Sphere Books Ltd., London, p. 43.

CHAPTER 3

Institutional Aspects of Integrated Planning Processes

Klaus Zimmermann

INTRODUCTION

THE discussion of problems concerning integrated planning[1] has predominantly been the domain of scientists, who have developed approaches toward the construction of comprehensive integrative models. The political impact and results from these models have remained amazingly low and, moreover, the few exceptions that exist do not significantly alter this negative perspective. This may lead to the general hypothesis that integrative approaches are not well-received within the political administrative sector and that such efforts are extremely difficult to promote against so many obstacles. A classic example of the problems that can arise in the implementation phase of integrative methods is illustrated in the planning-programming-budgeting system—an extremely comprehensive system, complicated by the fact that the budgeting process represents an integral part of it.

There are various reasons for the actual failure of integrative planning methods: there are, to name a few, the huge data requirements; the complexity of the problem and the method itself; the danger of a lack of discretion in political decisions; the distance from participatory involvement; and last but not least, ineffective institutional structures. These structures cannot be underestimated in their resistance to integrative planning methods. This is especially true when a high demand for allocative optimality is connected to the necessity for a new institutional delineation of regions or a new definition of actors. The idea of rapidly changing traditionally 'learned' institutions (with which people are familiar) to accord with the conditions for optimality, but without any additional costs, seems to be extremely short-sighted. As complex organizations, institutions cannot be viewed as systems of co-operative relations alone, since they also exist and function independently of their actual members. As a rule, institutions are directed to 'abstract persons:[2] The establishment of institutions as a substitute for spontaneous co-operation between certain individuals is part of the historical process of the development of the state and, therefore, institutions become especially rigid when confronted

[1]The author wishes to express his gratitude for the most valuable assistance given him in the preparation of this article by Prof. Dr. U. Simonis, Volkmar Hartje and Ms. Mary E. Kelley.

[2]This particular view stems from the political systems theory; see Easton (1965a, 1965b). Or, for an excellent piece of work in the context of decentralization, see Wittkaemper (1978).

with demands evolving from the growing and more complicated relations to their environment. This seems to be one of the essential reasons for the failure of integrative approaches, as far as implementation is intended. There is a lack of understanding with regard to the social benefits from traditional institutions and the costs of revising existing systems are habitually underestimated.

These initial remarks, based on political systems theory, may point out what the following approach cannot and will not achieve—namely, to design an optimal system of institutional organization of integrative planning. A discussion of the possible criteria for defining institutions for an effective integrative policy will be followed by an assessment of usefulness of such criteria, on the basis of the capability to realize and implement integrative approaches—two seemingly very different problems. The starting point of this discussion will focus on environmental policy within an integrative planning process and the institutional assignments will be exemplified as they relate to this policy area. The problem concerning criteria, however, have a general validity making them applicable to other policy areas as well.

THE STRUCTURE OF INTEGRATED PLANNING PROCEDURES (MODELS)

The common goal of every socio-economic development planning can be seen as the realization and maintenance of an equilibrium between the socio-economic and the ecological system.[3] To achieve that goal, it is necessary to identify and, hopefully, quantify the complex relationships and mutual effects of the various elements of socio-economic and ecological systems—all of this within a common framework. The starting point of this effort can be seen in the general classification of the activity sectors, production and consumption, both of these, in turn, influencing the environment via emissions of residuals and usage of raw materials. In addition, both activity sectors compete with the ecosystem for the use of scarce land. These relationships, and their respective feedbacks and repercussions, must be both acknowledged and incorporated into any discussion of political strategies for developing an equilibrium between the socio-economic and the environmental systems. When these interrelationships are known, the given resources potential and the planning goals become the starting points of the active planning process.

The term 'resource potential' is usually defined as the sum of the natural, produced (capital) or human resources available for production and consumption at a given time. For instance, water potential can be understood to be that amount of water which can be used by nature or by individuals, under the constraints of the given regeneration capacity. The available water potential is a function of time. As a further illustration within this context, the labour

[3]Comparative definitions of equilibria in the fields of economics and ecology have been provided in a stimulating book by Boulding (1978).

potential is a function of education, the societal fixed labour time, and so forth.

Therefore, for the planner, the received resource potential is represented by that amount of resources which may be used or consumed within a specific period of time; this represents a normative decision *sui generis*. In a simple system, one can distinguish between the socio-economic and the ecological potentials, which must then be pitted against the sum of those demands placed on them, resulting in the formation of an analogous structure of socio-economic and ecological elements on the demand side. Assuming, for example, that the regional policy goal is the increase of income, then a set of combinations of additional labour and capital factors, additional infrastructure, location and siting areas, etc., becomes necessary — that is, the fulfilment of the proposed goal for a region can be interpreted, within this system, as (additional) demand for the resource potentials.

Under the restrictions of the given potentials, the planner must decide how resources and activities within the region should be allocated. Performing this task accurately, however, requires a second step in the planning process: the analysis of the side-effects of those resource and activity allocations. This second step in the planning process becomes particularly important because it can indicate a necessity for revising the initial goals and demands on regional resources and perhaps even the initial assumptions concerning quantity and quality of the available regional development potential as well.

These complex interrelationships and procedures are often worked out in models; the structures, choice of model types, etc. need not be discussed here.[4] Actually, politically relevant models are only those which permit variation among the normative components in an interaction process between the planners and the politically responsible actors. A specific example of this type of procedure and model is illustrated in the Thoss-model for the Frankfurt region in the Federal Republic of Germany, where individual sub-models (for land use, waste disposal and recycling, sewage treatment, traffic, and energy) have been constructed.[5] The question of how the problem of 'regionalization' in a very heterogeneous region has been solved in this model is particularly interesting for the institutional aspects mentioned here. This seems to be an extremely difficult question, since all sectoral planning (i.e. every sub-model respectively) has a different perspective. Thus, the water-quality part of the model requires regions representing river basins; the solid-waste management requires a transportation model which provides connecting arteries between the primary locations where waste is generated and those where it is treated; further, the land-use model requires a segmentation based on natural units; and economic data can only be obtained with respect to the given political units and so on. For each individual environmental sub-model, this approach has resulted in singular delineations of regions, which are bound to vary in

[4]For an excellent discussion of different types of models, see Nijkamp (1977).

[5]There are many publications dealing with the Thoss-model. A very informative description of the model and results can be found in Brasse, Burgbacher and Wiik (1979).

accordance with changes in the data over time. A common characteristic shared by all such regional delineations is that they are only very rarely identical with the actual political units (cities, districts, etc.). Therefore, institutional congruence with existing political institutions is exceptional. Hence, the introduction and establishment of such *ad hoc* institutions will probably generate remarkable costs, one example being the learning costs for the population in becoming familiar with the new institutional structure. Moreover, such costs are normally not taken into account within the optimization-model framework, owing to obvious problems of quantification.

Theoretically, the evaluation of such institutional changes represents no specific problem. The criterion of optimal political interventions holds true here: that is, changes will take place as long as marginal social costs do not exceed the marginal social benefits. This criterion is a universal law. However, planning models in most previous cases have behaved as if the new optimal institutional structure could be achieved with zero transaction costs; this clearly seems to be an underestimation of the 'true' social costs of such corrections. Therefore, one most certainly has to admit the validity of the statement made by Fishelson (1979): "The optimal degree of federalism (regionalism) cannot be found by economic conventional (mathematical) optimization models. First, the formulation of the problem is non-feasible; second, many of its components are not quantifiable. Thus, common sense, assisted by economic facts and figures, is the tool for assigning political responsibilities and powers for improving environmental quantity."

CRITERIA FOR DEVELOPING THE INSTITUTIONAL PATTERN OF INTEGRATED PLANNING

On the basis of the result that comprehensive modelling will yield no further help in answering the institutional question, we must look for other, maybe less sophisticated methods for gaining additional insights into the institutional structure of integrative planning. The criteria to be discussed find their origins essentially in the economic theory of federalism. Contributions, primarily linked to the work of Mancur Olson in the framework of the externality/ internality concept, as well as the idea of fiscal equivalence, must be discussed at the outset (Olson, 1969 and 1979). If planning and the planning process are understood as a public good, then three cases concerning the regional extension of public goods and their respective administrative boundaries must be discussed:

(1) the public good can be used outside the administrative boundaries (externality);
(2) the public good reaches only a part of the population within the administrative boundaries (internality);

(3) the public good, and respectively, the regional extension thereof, coincides with the administrative boundaries and is financed by revenues from the regional institution (fiscal equivalence).

In the first case, externality, there is an obvious regional inefficiency to be noted: if the benefits of a public good partly accrue outside the administrative boundaries, the institution produces external economies. And consequently, this will lead to a supply of the public good which is below Pareto-optimality. Some classic examples of this externality problem can be seen, for instance, in the military expenditures of a country, which appear as external economies of its allies, or city/government expenditures for public goods that partially supply a larger metropolitan area.

The second case, internality, also suffers from allocative inefficiency. The public good reaches only a part of the population within the administrative boundaries, the consequence of which is, as before, the sub-optimal supply of public goods. Because the notion of internality is less familiar than its externality-counterpart, an illustration may serve to clarify this point: consider a central government with national taxes but no state and local jurisdiction. In a situation of of this type, coupled with a democratic voting system, the supply of collective goods for a local area will actually hurt more individuals than it will help. (This also holds for the case where Pareto-optimality would have required the production of this public good.) Therefore, if the benefits are local and the taxes national, the production of the public good leads directly to a majority of losers compared to the number of gainers—a valid result even if the supply of the public good indicates a cost/benefit relation not exceeding one. In such a case, it is efficient to establish a local taxing authority, even when the possibility for logrolling among such local entities exists. In general, logrolling is not very likely to occur because it is not very efficient among large groups of people. But even when logrolling does occur, there is still no certainty that there will be a proportional relation between the marginal costs and the marginal benefits, measured by the marginal rate of substitution of private and public goods and, hence, Pareto-optimality (Lindahl-equilibrium).

From these two opposing poles, of externality and internality, we are led directly to a third alternative concept: fiscal equivalence. There is a need for separate governmental instititutions representing each collective good with an unique boundary, such that a balance is established between those who received the benefits, on the one hand, and those who pay the costs, on the other. This is what Olson (1969) calls fiscal equivalence. This implies that there will be up to **x** separate institutions involved in the supply of **x** public goods. Consequently, the question of the usefulness of such organizational patterns arises, particularly when the integration of costs (organizational, bargaining, monitoring, and negotiating between planning levels) is taken into consideration. An institutional structure, in accordance with the principle of fiscal equivalence, would inevitably lead to a complex system of political and taxing responsibilities. The management of such a structure would be extremely difficult, in terms of both the administration and the people affected by it. Nevertheless, this does not alter the conclusion that the notion of fiscal equivalence is a very useful theoretical concept. Its political transportation, however, would lead to so many difficulties that, in practice, we shall have to give up the goal of Pareto-optimality in favour of weaker (singular) criteria of instititutional structure and integrated planning. Weaker criteria have already been proposed in an article by this author some years ago (Zimmermann, 1977). These were later discussed on the basis of environmental media in an expanded supplementary article by Ewringman and Hansmeyer (1979).

Like the problem of regional organization of a state in general—especially with respect to management in the realm of the provision of public goods—the problem of choice of optimal regional actors in environmental policy is intricately bound in various ways to political and legal aspects. If economic criteria alone cannot yield any conclusive answers to these problems, economic theory (in particular, the economic theory of federalism) can at least offer some hints for institutional assignments. Such criteria include:

(a) the regional preferences of the population with regard to certain environmental goods;
(b) the technical aspects of the provision and supply of these public goods; and
(c) interregional external effects.

In considering the importance of (a), differing regional preferences for certain environmental goods, one can plausibly claim that, although the preferences within a given region will never be totally homogeneous, the intraregional differences will normally be smaller than the interregional differences. This leads to the conclusion that the design of policy, in accordance with regional and political preferences, should generate different political concepts for the individual regions.

In considering the output of political processes alone, some hypotheses concerning the type and importance of regionally differing preferences, with regard to environmental protection, can be formulated as follows:

— When the significance of environmental deterioration increases for the population, their regional preferences become (more) homogeneous. For instance, when health hazards become a direct threat, we can expect uniform regulations and standard-setting throughout an entire region; and if this is a national problem, we can expect the same on a nation-wide basis.

— Regional preferences, on the other hand, will be more diverse, the more that specific environmental goods come to qualify as consumption goods. In particular, in those cases involving environmental goods linked to a high income-elasticity of demand (public luxury goods), differences in the evaluation of the benefits of environmental goods will correspond to the regional income distribution or to the distribution of incomes among different regions. One example illustrating this case would be the different estimations given for the recreational value of landscape or the question of desired water quality. Here, we can assume that uniform preferences for the whole state can only be expected with regard to the minimum standards and that regional differentiation will occur in accordance with higher degrees of water quality.

As long as differentiated regional preferences exist, a correspondingly differentiated allocation of authority at the regional level for environmental policy would be adequate—but only insofar as it relates to the luxury part of public goods.

The second criterion (b), for optimal regional decision-levels in economic policy, concentrates on the cost advantages of larger economic units — the 'economies of scale'. The per capita costs of supplying the population with public goods change in accordance with the size of the regional unit; the degree of these economies of scale depends on the technical qualifications of the public goods. Nearly all empirical studies on the cost curves of public goods yield downward sloping, or at least horizontal, cost curves related to units of output when the provisional area is enlarged. Progressive curves only occur very seldomly — a fact which confirms the general economies-of-scale hypothesis within the area of public goods.

As regards cost patterns in environmental policy, empirical evidence supports the claim that costs are decreasing in the areas of water provision, sewage systems and sewage treatment. Conversely, however, measures for the prevention of air and noise pollution have not been shown to yield a negative relation between provisional areas and cost curves. Such global results, however, cannot serve as a basis for decision-making; rather, they can only indicate tendencies which are subject to change as technology improves. In actual planning cases, the specific structure of production and settlements within the provisional areas must undergo additional analysis, as also must the degree of environmental deterioration and the regional population data. But information about cost curves is a necessary condition for evaluating the extent to which larger provisional areas enjoy cost advantages. In environmental policy, the largest economies of scale obviously exist for sewage treatment.

The third criterion (c) mentioned, for optimal organization and allocation of authority in environmental policy, is the criterion of maximization of external interregional effects—in other words, 'catching the spill-overs'. Here we return to the concept of fiscal equivalence, as previously discussed. Only when all costs and benefits of the activities within a region are internalized within its boundaries can an efficient provision of the public good (environment) be reasonably expected. But in practice, if complete internalization were a political goal, the design of regional units for most public and environmental goods would have to be so large that the positive welfare effects of regionalized organization (in accordance with the aggregation of preferences) would be diminished. Therefore, the criterion of 'catching the spill-overs' can only be employed as one of several criteria for determining optimal allocation of authority.

In attempting to sketch the optimal allocation of authority in environmental policy, the following structure seems to be plausible (table 1): On the basis of all three criteria mentioned, a local authority responsible for noise prevention, as well as conventional solid-waste management (garbage disposal), would appear to be adequate. Sewage treatment, however, would fall under the responsibility of regional authorities because of cost degression and high external effects. A regional authority would also have responsibility for air-pollution policy, where external effects are numerous but where there are no significant economies of scale. Further, since there will be higher ecnomies of scale in solid-waste management, accompanied by an increasing capital-intensity, a higher level of authority would also share responsibility in this sector of environmental policy.

Table 1: Institutional Assignment of Environmental Policy Areas

Environmetal policy areas Criteria of assingment	Noise prevention	Traditional solid-waste management (garbage collection and disposal	Water Sewage treatment	Air-pollution control	Modern solid-waste managm. (garbage incineration)	Health effects/ minimum standards of environmental quality	Facility siting/ consumptive use of nature (recreation, protection of nature and monuments)
Homogeneity of preferences	local	local				national	
Utilization of economies of scale	local	local	regional		regional		
Minimization of inter-regional external effects (spill-overs)	local	local	regional	regional			national

local
regional = resulting assignments corresponding to dominant criterion of assignment
national

With regard to authority at the state and national level, there are hardly any constraints arising from cost restrictions. State and national authorities will be optimal whenever there is a high degree of identity among regional preferences. This, for example, should be especially true in the case of health hazards and political decisions governing minimal standards of environmental quality. Cases representing greater difficulty, in terms of allocation authority at the state or national level, are those which result from significant interregional effects. One contemporary illustration for this is the selecting of sites for nuclear power plants. Here, planning with respect to relatively scarce, industrial-siting locations falls within the framework of national land-use policy. Further, in the consumption sphere, responsibility for such things as protection of recreational areas, natural and other monuments of historic value, etc., should be assigned to a higher (than regional) authority as well.

Normally, however, authority allocated on the basis of economic criteria will not be identical with the real authority allocated as a result of the historical process. So, if real authoritative status and optimal authoritative status diverge, resulting in a greater concentration of power at the local and regional levels, then from the standpoint of the 'spill-over' argument, the situation is difficult. On the other hand, when higher levels of authority (state, national) play a dominant role, the impact is far less important; here, at least theoretically, the possibility still exists for implementing policy on the basis of regional differentiation at the higher authoritative level. Nevertheless, this empirical argument will not negate those conclusions inferred from the 'internality' argument.

One particular problem concerning the allocation of authority stems from the relationships among cities within a large, highly urbanized region. Because of their close proximity to one another, the relatively high intraregional mobility and various economical and technical interconnections, it is reasonable to expect strong, interlocal, external effects through the provision and protection of environmental quality. Theoretically, in this situation it is possible for cities to bargain with one another directly; but there is a strong incentive for all of the cities to wait until just one of them provides the public good in question, thus allowing the others to go along for the ride at no cost to themselves. Opting for this 'free-rider' position is highly plausible in those cases where the public good is relatively indivisible, has strong external effects and can only be provided by facilities meeting certain minimal size-specifications. This situation seems to be especially important for big cities in an agglomeration, with many smaller cities in the surrounding area. It represents a typical example of 'exploitation of the great by the small', where the larger city is under pressure from demands so strong that the smaller ones can then go along for the ride free of charge (Olson, 1979, p. 182). The result of this is an efficiency disadvantage which normally cannot be corrected by direct negotiations between the cities concerned, but rather only through special types of co-ordination which, in the extreme, must be enforced from a higher federal level.

CRITIQUE OF THE PROPOSED CRITERIA

A good institutional structure is one which satisfies criteria stemming from legal, political, geographical, technical and natural resources considerations. The reason, why simple, productive results, in terms of an institutional fabric of integrated planning, have not as yet been realized is that even the most simple and elegant model contributed by one discipline becomes an extremely complex structure when aggregated over all the disciplines. More specifically, the criteria proposed by economists alone lack sufficient disciplinary breadth to cover such a comprehensive problem (Brown, 1979, p. 167). This fact provides a strong general argument in support of interdisciplinary efforts but, of course, does not imply that efforts by a certain individual discipline ought to be disregarded.

Turning to a critique of the criteria proposed, it can first be stated that, in practice, these criteria are generally too imprecise to solve the problem of institutional structures. For example, with respect to questions of environmental policy, the criterion of 'containment for all external effects' would result in a coverage which is far too extensive. Airsheds, for instance, are intimately connected to watersheds via substitution effects and technical relationships. Watersheds, in turn, flow into groundwater aquifers or spill into large water basins. Water from large basins flow into tidal estuaries and from these to the ocean. What is obviously needed then, concerning the criterion of catching or containing all externalities, is an effective trade-off process between increasing comprehensiveness and decreasing management feasibility.

The second criterion, 'homogeneity of preferences', also seems to be too imprecise for practical purposes. It is questionable, for example, whether preferences in a smaller geographic region are more homogeneous in reality than those in larger units. There is a great deal of evidence to support the contention that preferences are bound to resource-use groups, the members of which are often spread over distances of hundreds of miles. Therefore, this criterion stands on a relatively weak basis; it is only of limited use when conflicting resource-use groups are located in the same region.

Finally, studies on the effects of the size of institutions supplying public goods always seem to be useful. However, so long as the institutional costs and the costs of changing institutional structures are not covered, the 'economies of scale' argument is decisively unbalanced and only of limited value. Even famous theoretical contributions towards dissolution of these problems, within the context of interdependence costs (Tullock, 1969) and fixing an optimal degree of centralization, as the minimum of the vertical sum of spill-over costs and meritorization costs — none of these methods have, or will, yield practical results as long as empirical studies on the effects of the size of institutions are not available.

In summing up, it must be stated that, obviously, none of the criteria discussed are sufficiently well-defined and free of contradiction to serve as adequate instruments for integrated development planning. Were these criteria applied solely for the purpose of determining optimal solutions, they should at least be sufficient to indicate the direction of development within institutional structures. But this also seems not to be the case. For example, with reference to water-quality management, it is questionable whether such criteria can in fact account for changes within that institutional structure, even to some extent. The notion of 'fiscal equivalence' has its own merits. Its chief proponent, Mancur Olson, uncovered significant evidence indicating the development of a multi-level complex of jurisdictions in Europe (Olson, 1979, p. 184). In contrast to these observations, however, opposing results have been obtained. Gardner Brown, for example, points out the decrease in the number of water-management jurisdictions in Great Britain from 1,550 to 10, and in the Netherlands, from 1,200 to 29, for the period 1963 to 1978. His findings indicate a clear tendency towards comprehensive management and large management units (Brown, 1979, p. 170). The recognition that decentralized states, like the United States, tend to centralize while centralized states, like France, tend to decentralize shows only a very general trend towards convergence, further indicating that the optimal degree of centralization lies somewhere between both extremes.

Therefore, it must be concluded that the criteria discussed from the economic theory of federalism are capable only to a very limited extent of hinting at optimal solutions of regional management — a function far removed from serving as a basis for accurate predictin of power and responsibility within multi-level institutional structures. If there are no clear answers to questions concerning optimal institutional structures, or no clear explanations of historically grown institutions within a policy area, (like environmental policy), then finding solutions to the problem of integrated planning procedures represents an even more complicated exercise. What seems to be needed, therefore, is a theory that explains and predicts both the pace and direction of the trends discussed and within which the criteria discussed may represent significant elements.

CONCLUSIONS AND PROSPECTS

As we have seen, the explanatory power and the operationalization of these criteria will hardly be sufficient for designing an optimal institutional structure for integrative planning. Therefore, we shall have to lower our aspirations and try to formulate general rules for an institutional structure. A list of such general rules[6] might include the following:

(1) The better defined the general principles of policy at the federal level, and the stronger the delineation of goals and tasks between different institutions in the federal state, the higher will be the degree of federal decentralization possible.
(2) The most important rule governing federalism in an allocative view contains these requirements for the prevention or internalization of

[6]For an intense and comprehensive discussion of such possible rules, compare Frey (1977).

spill-overs. Those benefiting from public goods should be essentially identical with those bearing the costs and making the decisions regarding supply of public goods.

(3) Since the political boundaries do not necessarily coincide with the geographical distribution of the users of public goods and institutions, and since these normally cannot be made to fit within the boundaries defining the group of beneficiaries, the creation of special regions or districts would seem to be imperative.

(4) Since there are many difficulties associated with democratic decision-making and control within these artificial (i.e. not historically grown) structures, these special regions or districts should be designed only for the central infrastructure sectors.

(5) Interregional and regional burden-sharing techniques must be improved in order that ordinary political institutions, states and cities be able to supply those public goods having significant spill-over effects. This would seem to be a reasonable procedure for guaranteeing a functional, but not institutional, congruence of those benefiting from public goods and those bearing the costs for their supply.

(6) If a federal structure has been basically accepted, a certain range of discretion must be allowed, for regional and local institutions to autonomously select the sources and determine the size of their revenue.

(7) Multi-level planning (whether integrated or not) leads to an imprecise delineation of responsibilities. This, in turn, tends to (a) weaken incentives and sanctions for a good policy, (b) substitute political steering with bureaucratic steering, and (c) place restrictions on the necessity for fitting policy to the preferences of the population.

(8) Multi-level planning procedures, therefore, should only be accepted in those cases where consideration of regionally different preferences is either irrelevant or undesirable. In the Federal Republic of Germany, for example, for nearly all public tasks performed in accordance with multi-level planning procedures (Gemeinschaftsaufgabe), this conclusion is not respected.

These rules restrict an integrative planning process only insofar as they are related to multi-level planning procedures. In all other examples, there are possibilities for functional integration by using regional, burden-sharing techniques or by creating special districts with autonomous, taxing and spending authorities to supply public goods in central infrastructure sectors. But even here, it must be re-emphasized that the requirements for minimizing the distances between historically and politically learned institutional structures must be traded-off against the requirements for the optimal relevation of preferences.

The discussion of such complex issues is still only in its infancy and a good economic theory within this sector of analysis still seems to be lacking. The question therefore arises as to whether there are approaches other than economic ones that would produce better results. Moreover, it must also be asked whether an economic formulation of this problem actually implies doing the second step before the first one has been done.

An alternative approach, based on political science, would rest on the question of preconditions for usage of decentralized structures by the individual, which in turn would require mediation of various kinds of competence:[7]

(a) expert competence, in the sense of procurement of a certain amount of political expertise;
(b) integrative competence, in the sense of capacity to condense scattered insights into a general picture of political relevance;
(c) social competence, considered as the ability to realize and apply skills gained from expert and integrative competence in social activity, in participatory processes; and
(d) cultural competence, considered as the ability for self-realization via individualistic, political value-decisions.

If these competencies — and hence, the preconditions for the use of decentralized structures — are given, it is necessary to draw a distinction between different policy areas in accordance with the criterion of nearness to — or its correlate, distance from—acceptance. This is because not all policy areas are affected by the obligation of nearness to acceptance (capability for resolution via consensus) as are spatially linked, political activities. But even in the case of spatial phenomena, there seems to be no way for moving directly from the nearness-to-acceptance criterion to a call for establishing participation in this policy area; cost/benefit analysis of centralized and decentralized structures must necessarily be inserted at this stage. The following criteria could prove useful in such a cost/benefit analysis:

(a) **Operationalization of values:** The constitutions of the Western nations contain personalistic values which result from a national trade-off process between the poles of collectivism and individualism. Every organizational model of integrative planning, then, must test successfully for its ability to mediate constitutional decisions favouring particular collective values, to make them more visible for the citizen, so that insight into societal values will generate acceptance of the state order; in other words, legitimacy (that is, mediation of values as a contributing factor of their realization).
(b) **Optimization of responsibilities:** There are many horizontal and/or vertical relations between the various functions of government. Optimization of responsibility means finding those institutional assignments which guarantee the highest degree of persistence and integration in the implementation process, subject to the criteria

[7]These and the following arguments are based on the discussion in Wittkaemper (1978).

of efficiency and austerity, such that regulation of the collected responsibilities will be legitimate.

(c) **Optimization of participation:** The opportunity afforded by many constitutions to participate in public decision-making may lead to maximized participation. Such participation can be seen as a danger because, on the one hand, citizens are permitted to exercise their participatory rights in cases of strongly segmented and partial decisions; on the other hand, they may develop feelings of disappointment and alienation as a result of missing insights into general, overall relationships or as a result of a lack of participationin extremely sensitive areas of partial decision-making, such as the siting of nuclear power plants. Therefore, maximization of participation cannot be a societal goal; as such, the goal should be optimization of participation, whereby the opportunity is afforded to participate in substantial realizations of those government functions which mediate between insight and acceptance.

(d) **Social effectiveness:** Social effectiveness, as the yardstick for measuring problem-solving capabilities in the public sector, has two dimensions. The first relates to the internal organization of the administrative sector; the second relates to the social goals aimed at by the internal administrative process. Therefore, social efficiency at first affords the application of well-known criteria of management efficiency, such as profitability of capital, cost reduction or qualitative improvement. In the second stage, it is necessary to answer the question of how a given measure is to be judged in accordance with, firstly, the criterion of social opportunity costs of resources and secondly, the criterion of resulting social outputs, achievements and effects.

(e) **Activation of citizens:** Participatory political culture cannot grow, as previously indicated, as a result of organizational remedies alone. It must be bred into citizens via the mediation of the four competencies just discussed, through political education and step-by-step practice in accordance with educational progress. On this view, testing for citizen activation implies judging the fruitfulness of organizational solutions for mediating those competencies.

(e) **Homogeneity of the federal state and regional individuality:** The German constitution, for instance, prescribes, on the one hand, the value of federal homogeneity—in particular, the requirement of jurisdictional and economic unity (Art. 71, Abs. 2). On the other hand, however, it guarantees regional individuality in accordance with historical and cultural coherence (Art. 29, Abs. 1). Any possible organization, or organizational structure, is thus constrained in such a way that it may not violate either of these two constitutional principles. This implies that practical solutions must follow an 'intermediate course' between both principles.

From a political science point of view, these criteria would seem to be the most important for a cost/benefit analysis of centralized or decentralized models serving the functions of government. Moreover, these criteria may also be used as guidelines for the design of institutional structures of integrative policy process. But it is still questionable whether, and to what extent, these criteria are operational — that is, that they can serve as general yardsticks particularly in cases of doubt. In the final analysis, it must be admitted that the institutional question of the assignment of political responsibility and power represents a genuine political problem, which definitely cannot be solved on the basis of scientific criteria alone.

References

Boulding, K.E. (1978). *Ecodynamics. A New Theory of Societal Evolution,* Beverly Hills/London.

Brasse, P., Burgbacher, W. and Wiik, K. (1979). *Environmental and Regional Targets in a Common Framework. A Decision Model for the Lower Main Area.* Paper delivered at the MAB 11 Expert Panel, The Development and Application of Ecological Models in Urban and Regional Planning. Bad Homburg, FRG.

Brown, G. (1979). Criteria for An Appropriate Institutional Structure. In: H. Siebert, I. Walter and K. Zimmermann (eds.) *Regional Environmental Policy.* New York/London, pp. 166-167, 170.

Easton, D. (1965a). *A System Analysis of Political Life.* New York/London/ Sydney.

Easton, D. (1965b). *A Framework of Political Analysis.* Englewood Cliffs, N.J.

Ewringmann, D. and Hansmeyer, K.H. (1979). The Institutional Setting of Regional Environmental Policy. In: Siebert, I. Walter and K. Zimmermann (Eds.) *Regional Environmental Policy* New York/London, p. 152

Fishelson, G. (1979). Political and Jurisdictional Systems for Regional Environmental Policy. In: H. Siebert, I. Walter and K. Zimmermann (Eds.) *Regional Environmental Policy.* New York/London, pp. 171-173

Frey, R.L. (1977). *Zwischen Föderalismus und Zentralismus Ein volkswirtschaftliches Konzept des Schweizerischen Bundesstaats.* (Between Federalism and Centralism. An Economic Concept of the Swiss Federal State). Berne Frankfurt.

Nijkamp, P. (1977). *Theory and Application of Environmental Economics.* Amsterdam/ New York/Oxford.

Olson, M. (1969). *The Principle of Fiscal Equivalence: The Division of Responsibilities Among Different Levels of Government.* American Economic Review, Papers and Proceedings, Vol. 59, p. 479.

Olson, M. (1979). On Regional Pollution and Fiscal Equivalence. In: H. Siebert, I. Walter and K. Zimmermann (Eds.) *Regional Environmental Policy.* New York/London, pp. 181-182, 184.

Tullock, G. (1969). *Federalism: Problems of Scale.* Public Choice, Vol. 6, p. 19.

Wittkaemper, G.W. (1978). *Begründung einer stärkeren Dezentralisierung politisher Entscheidungen aus einer politologischen Theorie des Föderalismus.* (Reasons for Increasing Decentralization of Political Decisions within a Political Science Theory of Federalism). Paper presented at the meeting of the Konrad-Adenauer-Foundation on Alternative Concepts of Space-oriented Political Activity, Bonn.

Zimmermann, K. (1977). *Regionale Dimension in der Umweltpolitik.* (Regional Dimensions of Environmental Policy). Paper, Department of Landscape Planning, Technical University of Berlin.

CHAPTER 4

Interactive Learning Models for Integrated Planning

Peter Nijkamp and Jaap Spronk

RECENT TRENDS IN POLICY ANALYSIS

THE USE of programming models has a long history in the area of planning and policy analysis. Operations research methods especially have been widely applied in economic, environmental and energy policy problems, even on a global scale. The use of such models and techniques, however, also has been severely criticized due to the stringent assumptions underlying such advanced mathematical tools. The usual assumptions regarding mathematical models for policy analysis are:

(a) the existence of a clearly identifiable and unambiguous decision-maker or policy-maker;
(b) in a multi-person decision framework, the role of the successive decision-makers involved can be assessed precisely, either by defining an aggregate decision system or by estimating the relative power influence of the individuals or sub-groups (Blair, 1979; Saaty, 1977; and Shapley and Shubik, 1954);
(c) the objective(s) relevant for the planning problem concerned are exactly known (including their mutual trade-offs);
(d) spatial and social impacts of decisions to be made may be either neglected or assessed accurately via a spatial or social distributional systems model;
(e) equity and distribution problems (between either groups or regions) can be taken into account by means of the policy objectives and the structure of the special model in hand;
(f) the complex relationships between policy measures and policy objectives are precisely known via an operational economic model describing the various relevant impacts;
(g) the technical, institutional, social and economic side-conditions of the system concerned are also precisely known and can be specified in an operational way;
(h) the time trajectory of all variables of the system can be computed precisely;
(i) when the state of a system is characterized by uncertainty (for instance, due to stochastic variables), the probability distribution of the stochastic elements is known.

In the practice of decision-making, however, such conditions are hardly ever fulfilled, so that the optimal state of the system in hand is often an unrealistic concept. Consequently, traditional programming models tend to receive a fairly modest position in modern policy analysis. It turns out that the attention of policy makers has shifted from optimality analyses to impact analyses, effectiveness analyses and strategic-decision analyses. In such analyses, much more emphasis is placed on effects of policy measures, on shifts in social objectives and on conflict management and compromise principles.

Consequently, modern policy analyses have to be multidimensional in nature (Nijkamp, 1979; Nijkamp and Spronk, 1981; Rietveld, 1980) as they have to take into account the existence of a wide variety of social interests, decision groups and policy structures. This broader view on policy analysis requires an integrative framework for judging alternative policy options. Instead of designing optimizing systems, more emphasis has to be put on rationalizing systems by providing relevant information; revealing conflicts among objectives or groups; assessing trade-offs among different choice possibilities; gauging the distributive impacts of policy measures; identifying efficient (non-dominated) solutions; designing suitable and relevant methods for policy evaluation; and so forth. The current interest in interactive, multidimentional programming models for policy analysis clearly demonstrates the new trends in designing and employing formal tools for decision-making. This will be further discussed in the following sections.

POLICY ANALYSIS: CONTENTS

Given these recent trends in policy analysis, the following stages may be distinguished in setting up a policy analysis:

— the identification of policy objectives and of related judgement criteria for policy measures;
— the identification of all alternative choice possibilities which are considered to be relevant for the policy problem in question;
— the assessment of all foreseeable and expected effects of policy measures (or policy choices) upon the above mentioned objectives and criteria (for instance, by means of a formal structural model or an impact system);
— the identification of interest groups, and/or decision groups associated with the policy problem in question, as well as the assessment of conflicts among diverging priorities;
— the assessment of policy priorities and weights attached by policy-makers to the effects of measures taken by them;
— the development of appropriate evaluation methods and procedures (based on learning principles, for example);
— the treatment of information during the implementation stage of a

policy plan so as to get insight into the sensitivities of the policy impact analyses;
— an *ex post* evaluation of the actual policy decisions and of the role of the policy analysis in making these decisions.

Following on these remarks about policy analysis, criteria may be specified for a meaningful policy analysis:

— it should be able to assess the impacts of decisions or measures to be taken on policy objectives and/or criteria;
— it should provide a complete picture of relevant policy objectives, so that direct and indirect, intended and unintended impacts are included;
— it should reflect the variety and multidimensionality of the system concerned;
— it should be flexible, so that the policy analysis can easily be adjusted to new circumstances;
— it should be comprehensible to the decision-makers;
— it should be able to employ (available) data in an efficient way;
— beside efficiency criteria, it should take into account equity and/or other relevant social criteria;
— it should pay attention to conflicts among (interest) groups or other subsystems of the entire system;
— it should be able to assess trade-offs among different policy objectives;
— it should leave possibilities for a learning strategy in an (interactive) planning approach;
— it should be able to provide an integrated picture of all interactions and effects within the system in hand;
— it should open ways for compromise policies in case of policy conflicts;
— it should take into account the institutional structure of the existing policy framework;
— it should leave open a meaningful decision space for achieving a satisfactory state, based on either optimizer or satisficer principles.

MODELS IN POLICY ANALYSIS

A model provides a stylized picture of part of a complex reality. Clearly, models in a policy analysis should be able to indicate the boundaries within which policy decisions can be made, the trade-offs inherent in choosing alternative solutions, the impacts of policy measures on a (normally large) set of policy objectives and the possibilities for an interplay between experts and policy-makers.

Usually such a model is composed of a set of mathematical equations (Tinbergen, 1956) but this is not always necessary. Even impact systems and graph-theoretic representations might provide useful information (Blommestein and Nijkamp, 1981), while soft information can also be meaningfully taken into account (Nijkamp and Rietveld, 1981). Let us take the following formal model, containing a vector of decision variables z (instruments e.g.), of policy objectives $\underline{w}$ (with elements w_i, $i = 1, \ldots, I$), of endogenous variables $\underline{x}$ and of exogenous data $\underline{v}$:

$$\underline{f}(\underline{z}, \underline{w}, \underline{x}, \underline{v}) = \underline{0} \quad (1)$$

Then the following reduced form for the objectives may be assumed:

$$\underline{w} = \underline{g}(\underline{z}, \underline{v}) \quad (2)$$

Furthermore, a set of constraints (technical, social, political, economic, etc.) on the whole system may be defined:

$$\underline{z} \in K \quad (3)$$

where K represents a feasible area. Then, an efficient (non-dominated or Pareto-optimal) solution may be defined as follows: $z \in K$ is efficient, if no $\underline{z}^* \in K$ does exist, such that:

$$\underline{w}^* = \underline{g}(\underline{z}^*, \underline{v}) \geq \underline{w} \quad (4)$$

and

$$w_i^* = g_i(\underline{z}^*, \underline{v}) > w_i \qquad i \in \{1, \ldots, I\} \quad (5)$$

Thus, an efficient solution supposes that no other feasible policy exists, which is for all policy criteria at least equally good and for at least one criterion better (Despontin, 1980). Normally, one may expect that any good policy should be an efficient solution, although sometimes—due to political reasons or uncertainties—non-efficient solutions are also being selected (Leibenstein, 1976).

Therefore, a meaningful policy analysis should focus attention in particular on the efficiency frontier (i.e. the set of efficient solutions) in order to identify a policy that will not be dominated by other policies. This is especially important in the framework of interactive policy models which usually aim at finding a compromise solution located on the efficiency frontier.

OBJECTIVES IN POLICY ANALYSIS

The previous discussion was based on the assumption that policy objectives can easily be indentified and are given prior to the actual use of the model. In reality, however, neither the analysts nor the decision-makers have a perfect insight into the various objectives to be considered in a policy analysis. Clearly, official reports and documents of decision-makers may give some indications concerning objectives but these are often defined in a fuzzy way and left open to many interpretations when general objectives have to be translated into operational policy criteria. Moreover, during the process of policy analysis itself, new insights are obtained which may lead to a re-orientation and re-specification of policy criteria. Of course, it might be possible to include policy objectives as 'hard' constraints but this runs the risk of excluding policy flexibility from the model. Consequently, it is recommended to include objectives, wherever possible, in the form of objective functions instead of contraints.

The choice of objective functions should evidently be based on the priorities of decision-makers but should also be co-determined by the interests of other groups, including conflicts among objectives: conflict analysis is an essential ingredient of policy analysis. Furthermore, whenever possible, the policy objectives taken into consideration should not only refer to traditional welfare indicators (such as income or employment), but should also pay attention to 'soft' social or environmental indicators, so that the policy analysis in hand is based on a broad and balanced spectrum of policy considerations. In this respect, a policy analysis may also contribute substantially to gaining more insight into the political feasibility of compromise solutions, especially in the framework of an interactive policy approach with multiple objectives.

INTERACTIVE POLICY MODELS

Many problems in a policy analysis do not require an unambiguous solution that represents once and for all the optimal state of the system concerned. In light of the process character of many decision problems, an interactive policy analysis is, however, a reasonable approach. This approach is usually composed of a series of steps based on a systematic exchange of information between decision-makers and analysts. These interactive approaches normally have two common steps:

(a) the analysts propose meaningful and feasible trial solutions on the basis of a well-defined compromise procedure;
(b) the decision-makers respond to each trial solution by indicating in which respect (i.e. in regard to which effects) the proposed compromise is still unsatisfactory.

These steps can be successively repeated until, after a series of steps, a final satisfactory compromise solution has been identified. Recently, a large number of interactive models has been developed (Rietveld, 1980; Spronk, 1981).

Such interactive policy models, which have already demonstrated their usefulness on several occasions, have many significant advantages compared to traditional optimization methods:

— they are in agreement with the process character of many planning problems;
— they are built on learning principles for decision-makers;
— they provide necessary and meaningful information in a systematic stepwise way;
— they take into account the limited capability of the human mind to judge complex decision problems, with many choice options, in one step;
— they emphasize the active role of decision-makers in specifying and solving choice problems *inter alia* by making policy objectives more explicit;
— they are able to take account of the variety and conflicting nature of policy options or decision criteria;
— they allow an assessment of (implicit or explicit) trade-off in many choice situations, without necessarily requiring the specification of weights;
— they provide an integrative framework for choosing consistent compromise solutions;
— they may be used to successively eliminate less relevant alternative choice options;
— they may fit into an institutional structure based on multiple decision-makers or various decision levels.

In conclusion, interactive policy models may provide a coherent, operational and systematic contribution to a scientific rationalization of complex policy problems in reality. More explicit attention will now be devoted to one of the recently developed, interactive policy models, namely interactive multiple goal programming.

INTERACTIVE MULTIPLE GOAL PROGRAMMING (IMGP)

Consider the following multiple goal programming problem:

$$\left.\begin{array}{ll} \max g_i(\underline{x}) & \forall i \\ \text{subject to } A\underline{x} \geq \underline{b} & \end{array}\right| \qquad (6)$$

where $\underline{x}$ is a vector of instrumental variables (see also equation 2). One of the methods for dealing with multiple goal functions is interactive multiple goal programming (IMGP) (Nijkamp and Spronk, 1980; Spronk, 1981).

In IMGP, the decision-maker provides information about his preferences on the basis of a provisional solution and a potency matrix presented to him. The potency matrix consists of two vectors, representing respectively the pessimistic and the ideal solution. For each of the goal variables separately, the pessimistic solution represents a minimum value (usually proposed by the decision-maker), whereas the ideal solution represents the individual maximum values, given the pessimistic solution. The decision-maker has to indicate whether or not a solution is satisfatory and, if not, which of the minimum goal values should be increased in value. Then a new solution is presented to him, together with a new potency matrix. The decision-maker has to indicate whether the shifts in the proposed solution are outweighted by the corresponding shifts in the potency matrix. If not, a new solution is calculated and so forth.

For ease of presentation, the method is described here by assuming that in each iteration, one and only one element of the solution will alter. However, a generalization to more elements is straightforward.

Step 1

Maximize each goal variable g_i (x) separately, and denote the maxima by $\underline{g}^*_i$ and the I resulting values of the instrumental variables by x^{i*}, i = 1, . . . , I. It is not possible to find a feasible value of $g_i(\underline{x})$ that exceeds g_i^*. Generally, it is not necessary to accept a value of $g_i(\underline{x})$ which is lower than g_i^{min}, defined as:

$$g_i^{min} = \min \left| g_i(\underline{x}^{j*}) \right| \quad , j = 1, \ldots, I \qquad (7)$$

This is the lowest value of $g_i(\underline{x})$ resulting from the successive maximization of the goal variables. Then, the final solution S* must be found between the 'ideal' (but normally unfeasible) solution I, and the 'pessimistic' solution Q, which are defined respectively as:

$$I = [g_1^*, g_2^*, \ldots, g_I^*] \text{ and}$$
$$Q = [g_1^{min}, g_2^{min}, \ldots, g^{min}_I] \tag{8}$$

To facilitate the notation, we have included the ideal solution I and the pessimistic solution Q in the (2 × I) 'potency matrix' P.

Step 2

Define for all $j = 1, \ldots, I$

$$\delta_j = g^*_j - g_j^{min} \tag{9}$$

Step 3

Define the initial solution as:

$$S_1 = [g_1^{min}, g_2^{min}, \ldots, g_I^{min}] \tag{10}$$

which is thus equal to the pessimistic solution defined in (8). Present this solution together with the potency matrix P_1 to the decision-maker.

Step 4

If the proposed solution satisfies the decision-maker, one may terminate; if not define R_i as the subset of R defined by the goal levels in S_i, and proceed to step 5.

Step 5

The decision-maker has to answer the question: 'Given the provisional solution S_i, which goal variable should first be improved?'.

Step 6

Assume that the decision-maker wants to augment the j-th goal variable. Then construct a new trial solution $\hat{S}_{i+1}$, which differs from S_i only in the value of the j-th goal variable (denoted by $g_j(\underline{x})\hat{S}_{i+1}$ and $g_j(\underline{x})S_i$ respectively). Define next:

$$g_j(x)\hat{S}_{i+1} = g_j(x) S_i + \tfrac{1}{2}\delta_j \tag{11}$$

and introduce the restriction:

$$g_j(x) > g_j(x)\hat{S}_{i+1} \tag{12}$$

Step 7

Combine the restriction formulated in step 6 or in step 9 with the set of restrictions describing the feasible region R_i. Calculate a new potency matrix, like in step 2, but subject to the new set of restrictions. Denote this potency matrix by $\hat{P}_{i+1}$.

Step 8

Confront the decision-maker with S_i and $\hat{S}_{i+1}$ on the one hand, and with P_i and $\hat{P}_i + {}_1$ on the other hand. The shifts in the potency matrix can be viewed as a 'sacrifice' trade-off for reaching the proposed solution. If the decision-maker judges this sacrifice to be reasonable, accept the proposed solution by putting $S_{i+1} = \hat{S}_{i+1}$ and $\hat{P}_{i+1} = \hat{P}_{i+1}$ and put $\delta_j = \frac{1}{2}\delta_j$. Continue with step 4. If the decision-maker regards the sacrifice as unjustified, the proposed value of $g_j(x)$ is obviously too high. In that case, one may drop the constraint added in step 7.

Step 9

We know that $g_j(x)_{S_i}$ is too low and that $g_j(x)_{\hat{S}_{i+1}}$ is too high in the decision-maker's view. Set δ_j equal to the difference between these two values. Then a new proposal value $\hat{S}_{i+1}$ is calculated according to (11). Like in step 6, we add the restriction that $g_j(x)$ must equal or exceed the new proposal value and go to step 7.

A flowchart of the IMG procedure is given in figure 1. The method is entirely operational and has been used in various real-world problems.[1] For further details on IMGP (inclusion of aspiration levels, etc.), we refer the reader to Nijkamp and Spronk (1980) and Spronk (1981).

Figure 1. Simplified flowchart of the interactive framework

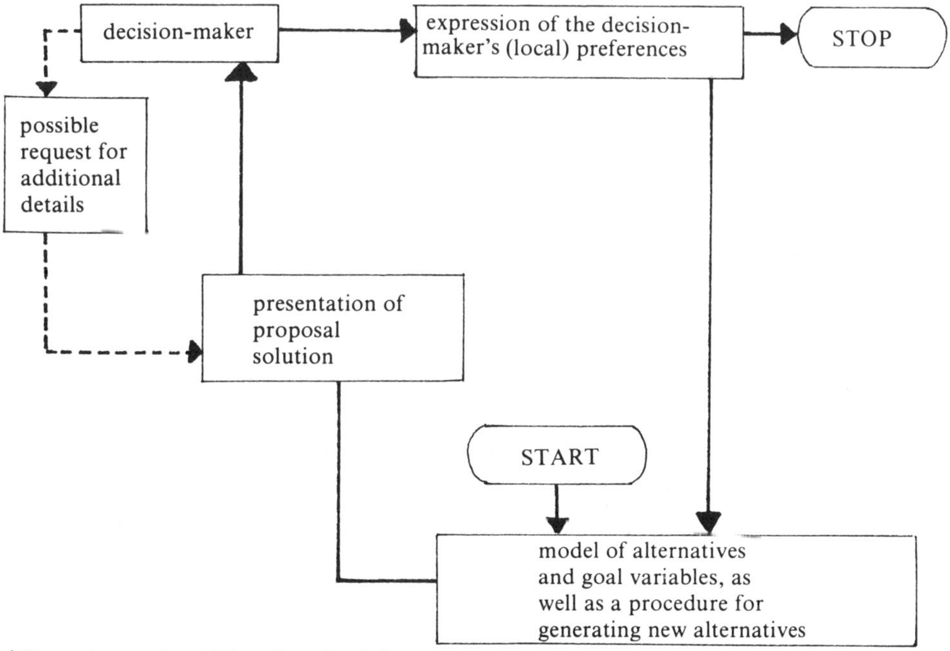

[1]The authors acknowledge the stimulating co-operation with Professor J. A. Hartog during the computerization and implementation phases of the interactive model.

ILLUSTRATION: A MODEL FOR WESTERN EUROPE

The model used in this empirical illustration was designed for a major industrial heartland in Western Europe, composed of the areas: Netherlands, Belgium, Nordrhein-Westfalen and France Nord, by van Driel *et al.* (1980). This part of Western Europe contains approximately 45 million inhabitants in an area of 115,000 square kilometres. In this industrialized and densely populated area, various conflicting options regarding economic growth, environmental conditions and energy availability are most likely to emerge.

Van Driel *et al.* (1980) have developed a dynamic economic sectoral model based on an input-output framework. The number of sectors in this model was 17. The structure of this input-output model, formulated as an inequality condition, is:

$$\underline{q}_t \geq {}_M (A+D)\, \underline{q}_t + K\, (\underline{w}_{t+1} - \underline{w}_t) + \underline{v}_t \tag{1}$$

with:

$\underline{q}_t$	:	vector of sectoral production levels in year t
$\underline{w}_t$	:	productive capacity in year t
$\underline{v}_t$	:	final demand and export surplus in year t
A	:	matrix of input-output coefficients
D	:	matrix with sectoral depreciation coefficients
K	:	matrix of sectoral capital coefficients

Clearly, the following condition holds:

$$\underline{q}_t \leq \underline{w}_t \tag{2}$$

This model was extended with both a pollution model, describing the emission of four kinds of pollutants (waste water, sulphur dioxide, solid waste and pollution from cars), and corresponding pollution abatement technology.[2] The dynamic input-output model was completed by means of capital and depreciation coefficients based on a vintage model. The model does not contain behavioural equations but, instead, policy options have been formulated in the form of inequalities, for example, regarding productive capacity (i.e. the capital stock) and environmental pollution. The planning horizon of the model was set as 10 years. After the specification of an objective function, this model can be treated as either a year-to-year programming model or a 10-year model. This has been used as a framework for an interactive policy analysis.

[2]In fact, the pollution coefficients for a sector were expressed as the abatement costs per unit of production value of the sector concerned.

The experiments with this interactive model were induced by the Dutch Scientific Council for Public Policy, abbreviated as WRR (WRR, 1980). Consequently, the Interactive Multiple Goal Programming approach was chosen as a tool to obtain more insight into the conflicting nature of different policy objectives, the feasibility of certain economic policy scenarios and the (in)stability of the results of the model for alternative policy variants.

After a discussion, six goal variables were ultimately selected for an interactive policy analysis. These were:

(1) Employment : maximization of the total wage sum over the planning horizon;

(2) Growth : minimization of the difference between the actual growth rate of production and a three per cent target growth rate of production;

(3) Environmental Quality : minimization of pollution by introducing a desired negative growth path for each pollutant (varying from five to ten per cent);

(4) Balance-of-payments : minimization of the maximum change in export surplus compared to a base year for each sector separately;

(5) Overall equilibrium on the balance-of-payments : minimization of the total deficit over the entire planning period;

(6) Stability of consumption pattern : minimization of the maximum annual decrease (or maximization of the minimum increase) for each sector.

On the basis of this set of six objective functions, a series of experiments with interactive policy strategies has been carried out in the framework of the WRR study, some results of which will be described next.

EXPERIMENTS WITH THE INTERACTIVE POLICY MODEL

Multiple-criteria-decision methods have been employed as useful tools in various integrated planning models (Despotin, 1980; Hartog *et al.* 1980). Recently, experiments with a small-scale version of the model just described were carried out. A fully operational version of this interactive policy model was demonstrated to various experts and decision-makers in government and industry. Two main conclusions can be drawn from these demonstrations: first, both the experts and the decision-makers regarded the interactive framework as a useful decision aid, mainly because of the induced learning effects; secondly, since the input-output model as proposed by van Driel *et al.* (1980) does not include any behavioural relationship, the results depend on the technical relations only and on the decision-maker's evaluations. These technical relationships are rather 'hard', so that one may conclude that, if a certain combination of the objectives is not feasible within the model, then this combination is certainly not feasible in the real world.

On the other hand, if a certain combination of objectives turns out to be feasible within the model, it is not certain—because of the omitted behavioural relationships—that this combination can be realized in practice. In summary, the results obtained by means of this interactive policy model are rather hard (although stated in negative ('falsification') terms). Further details of this small version of the model are given by Hartog *et al* (1980).

On the basis of these experiments, the Dutch Scientific Council for Public Policy (WRR) decided to implement the IMGP procedure in combination with the complete version of the Western Europe model, previously described. Because this project had to be carried out within a limited time period and because the experiences with the interactive procedure were not yet related to large-scale models, the results of the experiments were not as satisfactory as they could have been (Hartog and Spronk, 1980; WRR, 1980). Nevertheless, the results were judged to be satisfactory. Several conclusions were also drawn which turned out to be helpful in new experiments with interactive modelling (for instance, a new study of the WRR uses an adapted version of this methodology).

In table 1, the respective sets of limit goal values ('pessimistic' solutions), subject to which each goal variable had to be optimized separately at each iteration, are given. All figures are in billions of Dutch guilders (1965), except those relating to the third goal variable which are expressed in percentages. The first iteration consists of the computation of the unconditional optimal solutions. Thus, in the first iteration, each goal variable is optimized separately—without any constraint on the values of the (other) goal variables. From the second iteration onwards, the decision-maker can add minimally (or maximally) required (pessimistic) goal values. Analysis of the optimal solutions of the first iteration may be helpful for finding suitable pessimistic values for the second iteration. Until the fifth iteration, no pessimistic goal value is given for goal variable 6 because, until this very iteration, it was provisionally (but incorrectly) assumed that the balance-of-payment deficit could be decreased implicitly by means of the other goal variables. In fact, the realization of this mistake can be considered to be one of the learning effects obtained by using the interactive policy model.

More interesting than the pessimistic goal values are the optimal values which can be obtained while taking account of the pessimistic goal values.

Table 1: Sets of limits on the Goal Variables

Goal Variable		1	2	3	4	5	6	7	8	9	10
Employment	1	-	300	350	350	350	350	350	350	350	350
Deviation from growth target	2	-	19	10	10	10	10	10	10	10	10
Deviation from pollution decrease target	3	-	50	25	25	25	15	5	5	5	5
Maximum change in sectoral export-surplus	4	-	16	5	1	1	1	1	1	1	1
Maximum decrease of sectoral consumption	5	-	10	5	5	5	5	5	0.1	5	0.1
Total balance-of-payments deficit	6	-	-	-	-	10	10	10	10	5	5

Three examples are given in table 2. The set of goal values A is obtained when the first goal variable is maximized in the 10th iteration subject to the corresponding set of limits given in table 1. The set of goal values B is obtained when the 6th goal variable is minimized in iteration 7 and set C is obtained when the 3rd goal variable is minimized in the 7th iteration.

Table 2: Some illustrative outcomes

Goal Variable	A	B	C
1	513	439	350
2	10	10	10
3	5	5	4.6
4	1	1	1
5	5	10	0
6	0.1	0.33	0.23

For a much more detailed report of these results we refer the reader to WRR (1980). It should be stressed here that, within the pessimistic goal values in iteration 10 still many alternatives are feasible. These remaining alternatives were not studied in detail. This was partly due to the limited time available for the study. However, a much more important reason was that the obtained results had given new insights to both experts and decision-makers. It was felt that new restrictions should be added to the model and that some of the goal variables should be re-formulated. For example, the outcomes suggested that different formulations of goal variables, which are defined in terms of growth paths, may give rise to very different results. Furthermore, within a 10-year-period policy model incorporating many sectors, the number of goal variables tends to become unmanageable. Therefore, some of the goal variables were formulated in a minimax sense, i.e. as the maximum deviation (to be minimized) from a target growth path. In principle, all these deviations may be scaled in different dimensions. Thus, many different ways exist to handle the problem of large numbers of goal variables, not yet studied in full detail.

CONCLUSIONS

From the experiments described, several important lessons can be drawn. The main conclusion is that interactive policy models can be an important decision aid. It should be stressed that the primary purpose of these models is not to provide 'good', or even 'optimal', solutions (although formally speaking, this kind of solution can be provided by these models). The main use of these models is that they can be an important learning tool, both for decision-makers and experts. This is especially true since, in practice, exact goal definitions very often do not exist. Interactive policy models can help to define the decision-makers' goals. Furthermore, it is not always easy to identify the decision-maker(s). In this case, interactive policy models can serve as a flexible means of communication.

References

Blair, P.D. (1979). *Multiple Regional Energy Planning.* Kluwer Nijhoff, Boston.

Blommestein, H.J. and Nijkamp P., (1981). *Soft Spatial Econometric Causality Models.* Research Memorandum, Dept. of Economics, Free University, Amsterdam.

Despontin, M. (1980). *Kwantitatieve Economische Politiek vanuit een Besluitvormings-optiek.* Ph.D., Free University, Brussels.

Driel, G.J. van., Hartog J.A., and Ravenzwaaij, C. van. (1980). *Limits to the Welfare State.* Martinus Nijhoff, Boston/The Hague.

Hartog, J.A. and Spronk J. (1980). *Een Modelstudie naar de Relatie Milieu-economie.* Werkdocument WRR, Staatsuitgeverij, Den Haag.

Hartog, J.A., Nijkamp P., and Spronk J. (1980). *Operational Multiple Goal Models for Large Economic Environmental Systems.* Proceedings for the IX-th IFIP Conference on Optimization Techniques, Vol. II, K, Iracki *et al.*

Leibenstein, H. (1976). *Beyond Economic Man.* Harvard University Press, Cambridge.

Nijkamp, P. (1979). *Multidimensional Spatial Data and Decision Analysis,* John Wiley and Sons, Inc., New York/Chichester.

Nijkamp, P. and Rietveld, P. (1981). *Ordinal Multivariate Analysis.* Professional Paper 81-2, IIASA, Laxenburge.

Nijkamp, P. and Spronk J. (1980). Interactive Multiple Goal Programming : An Evaluation and Results. In *Multiple Criteria Decision Making: Theory and Application.* G. Fandel and T. Gal (Eds.) Springer, Berlin, pp. 278-293.

Nijkamp, P. and Spronk J. (1981). *Multicriteria Analysis: Operational Methods.* Gower, Farnborough.

Rietveld, P. (1980). *Multiple Objective Decision Methods and Regional Planning.* North-Holland Publ. Co., Amsterdam.

Saaty, T.L. (1977). *Scaling for Priorities in Hierarchical Structures.* Journal of Mathematical Psychology, Vol. 15, No. 3, pp. 234-281.

Shapley, L.S. and Shubik, M. (1954). *A Method for Evaluating the Distribution of Power in a Committee System,* American Political Science Review, Vol. 48, pp. 787-793.

Spronk, J., (1981). *Interactive Multiple Goal Programme Applications to Financial Planning.* Kluwer Nijhoff, Boston.

Tinbergen, J. (1956). *Economic Policy: Principles and Design.* North-Holland Publishing Co., Amsterdam.

WRR (1980). *Plaats en Toekomst van de Nederlandse Industrie.* Staatsuitgeverij, Den Haag.

CHAPTER 5

Public Participation in Environmental Planning in the USA

Dorothy Nelkin

"SCIENCE FINDS—Industry Applies—Man Conforms." This was the theme of the Chicago World's Fair in 1933, held to celebrate a century of scientific and technological progress. These relationships remained largely unquestioned during the period of unprecedented economic growth that followed the World War II. Such optimistic faith in progress, however, began to wane under the pressure of various social crises of the 1960s. In particular, the environmental movement brought growing public concern over the impacts of technology, disillusionment with the implications of conforming to technical imperatives and compelling pressures for more deliberate policies of regulation. Subsequently, the relationship between technology and the public has been marked less by conformity than by the search for control.

A significant aspect of this changing relationship has been the concern about public participation. The simplest definition of democracy, 'rule by the people', implies participation. To rule carries the unstated condition 'to take part'; yet the role of participation in democracy has never been agreed upon. To conclude from its etymology that democracy implies participation by the people in the activity of ruling begins, rather than ends, the controversy (Keim, 1975). Local citizens seek a voice in decisions involving the location and design of power plants, industries and airports. Environmental groups seek to influence governmental rule-making and standard-setting. Public-interest groups challenge the legitimacy of responsible authorities and the role of expertise in technical decisions. "Demystification", "accountability" and "participation" are the slogans of recent protests.

This challenge has stimulated a variety of administrative and legislative efforts to make political systems more responsive by involving citizens in the formulation and implementation of environmental policy. But these efforts raise a set of persistent and controversial questions: Who should participate in major policy decisions affecting the environment? What is the appriopriate role of expertise? How is it possible to include those affected by environmental policy choices in the process of decision-making in complex technical areas? As administrative bureaucracies grope for ways to include citizen-feedback in their decision-making procedures, they do so with great ambivalence. They fear that greater public participation may threaten future economic development. They struggle with the problems of defining 'the public' and of establishing appropriate goals when they tend to shift as more people are involved in decision-making procedures. They anticipate a deluge of demands by non-represented private interests or by self-appointed environmental elites. And they fear that this will significantly undermine administrative efficiency and autonomy. Indeed, the debate over the use of natural resources is paralleled by a debate over the extent and form of public participation appropriate in environmental decisions. This chapter explores the sources of participatory demands, the governmental reforms to extend participation in natural-resources decisions and their limits.

SOURCES OF PARTICIPATORY DEMANDS

Nearly every technological development, from the siting of nuclear power plants to the implementation of new technology-based programmes (such as fluoridation), confronts public opposition (Nelkin, 1971, 1975a). Technologies once welcomed as signs of progress are now associated with the degradation of natural resources. Critics see the negative impacts of science and technology as evidence that government and its experts cannot deal effectively with evolving social problems (Nelson, 1977). And they often attribute such policy failures to greater responsiveness to corporate industrial interests than to public needs. Thus, as critics engage in disputes over specific projects, they also begin to question the decision-making process. Concerns about substance evolve into questions about the legitimacy of authority and expertise and to demands for greater public participation in social-policy development and implementation.

The changing image of technology and the impulse towards greater public involvement are, in part, reactions to the encompassing role of the federal government. The rate of technological change, the scale of scientific and technical projects, the rapid diffusion of technology to new areas—all have brought extensive federal commitment to the support of science and technology. This burgeoning commitment has led to the growth of federal administrative agencies, placing technology more and more in the public sector (Galbraith, 1967). Demands for accountability, stimulated by the general loss of confidence in established institutions, follow from this changing relationship.

Disputes among experts about the environmental risks associated with technologies have further eroded confidence. As experts support the legitimacy of contending claims and raise doubts about the evidence backing policy choices, they expose the value-laden political dimensions of these choices. Disagreement among scientists confirms public fears about biased decisions and direct attention to decision-making procedures that are unresponsive to public needs (Mazur, 1973; Nelkin, 1975b). Examples are numerous: the controversies over the extent of eutrophication caused by nuclear power plants, over the impact of the Tocks Island Dam, over the relative costs and benefits of pesticides—all illustrate the erosion of confidence that follows technical disputes.

The translation of such public concerns into demands for direct participation draws on the experience and ideology of the 1960's War on Poverty, when the inability of elected officials and representative bodies to deal with the problems of poverty led to the social experiments labelled 'maximum feasible participation'. These experiments in community control reinvented the 'public' as a direct actor in social-policy decisions. Participation was perceived as a 'right', a means to reduce alienation among the powerless, a way to assure that government decisions reflect the preferences of the people (Cook and Morgan, 1971; Spiegel, 1968). The experience with participation in the 1960s was a confused and often disillusioning one; many of the experiments failed in their goals of redressing fundamental inequities or of enhancing trust. Perhaps their greatest impact, however, was the revival of the notion of direct participation, the growth of public support of the legitimacy of citizens' involvement and the extension of the democratic ethos into an ever-broadening set of issues (Marris and Rein, 1967; Yin *et al.,* 1973). This is certainly reflected in the environmental area.[7]

In part, it is the nature of the resources problem that calls for a participatory response. Environmental protection requires collective social responsibility in situations where an individual may perceive no personal compensation. "Virtue may be its own reward, but the reward is too often a collective good shared only minutely by the virtuous individual." (Schilling, 1974). Indeed, a major problem of environmental policy is how to enhance the individual's stake in the collective interest. Public involvement is one way to develop this stake. In addition, the need to enhance the legitimacy of decision-making procedures has shaped administrative and legislative efforts to involve citizens more directly in the formulation and implementation of policies concerning natural resources.

PARTICIPATORY REFORMS

During the late 1960s, demands for greater public participation from environmental groups coincided with Congressional qualms about the increasing power of administrative agencies. The major environmental legislation around 1970 — (the Airport and Airways Development Act, the Federal Water Pollution Control Act, the Coastal Zone Management Act, the Highway Safety Act, the National Environmental Policy Act, the Energy Re-organization Act)—all contained imperatives for the greater involvement of citizens. Because administrative agencies are a major arena of political action, often taking the place of legislative bodies in defining important policy problems and creating their solutions, they have been the main target for participatory reforms.

Examples of participatory reforms in the administrative process suggest both the variety of existing mechanisms available in the natural resources area and their limits. Most common are procedures which inform the public about government decisions. Other procedures channel information from the public to the decision-making agency. Finally, some procedures seek to open the administrative process to negotiation and compromise, whereby representatives actually share decision-making authority. These include citizen litigation and referenda; major channels for participation and public influence that are used in lieu of more direct possibilities for involvement in the administrative process.

Informing the Public

A major and important thrust of participatory reform is to provide public access to the data and reports that reveal the actual process of decision-making and the information on which decisions are based. The budget of an administrative agency is an important point of entry for citizens into decision-making, for it provides a tangible statement of priorities and intentions and helps to distinguish real operational choices from mere public statements. Detailed budget information is usually hidden from public view until decisions are made and compromises crystallized. However, some access to the federal budget process was provided when the Office of Management and Budget announced in Circular A-95 that all local applications for federal grants and assistance must be first reviewed at the state level. Local review, facilitated by regional clearing-houses for documents, has provided an opportunity for alert citizens' groups to study budget proposals before they have acquired governmental support.

The rules and standards that form the basis for regulatory action are another crucial type of information. The Administrative Procedures Act of 1948 provided that all rules and regulations must be published in the Federal Register prior to final adoption (USC, 1946). The Federal Register, like many government documents, is cumbersome, containing more than 60,000, three-columned pages per year. It is, however, widely available and contains the details of agency actions during the planning stages. Citizens' groups often scan the Register for reports relevant to their interests.

In 1966, the Freedom of Information Act extended public access to government reports and position papers (USC, 1975). This act requires all federal agencies to describe, in the Federal Register, their organizational structure and decision-making procedures. Agencies must make available on request policy papers, staff manuals and final opinions. Administrative agencies opposed the passage of this Act; they feared that increased exposure would inhibit frank, internal discussion of controversial issues. Administrative autonomy, it was argued, guaranteed more consistent, far-sighted and efficient decisions. Opposition delayed passage of the Act for eleven years, adding several loopholes that provide agencies with ways to provide compliance. Moreover, the high cost of obtaining information under the Act (agencies may charge search and copy fees) often discourages citizens' groups; most requests come from government employees seeking access to personnel records, businesses seeking information about competitors and lawyers involved in liability suits (Kolata, 1975).

Nevertheless, information released by this legislation has been important in several environmental disputes. Public awareness of pesticide toxicity followed from use of the Act. Judge David Bazelon, underscoring the importance of informing the public about technologies that "threaten to outstrip our collective ability to control its affects on our lives", used the Freedom of Information Act to release a government study critical of the SST (Wade, 1972).

The most significant advance in public access to information has been the requirement that all federal agencies prepare Environmental Impact Statements (EIS) "on proposals for legislation and other major federal actions significantly affecting the quality of the human environment" (CEQ, 1973). As EIS must include a detailed appraisal of plausible environmental damage, feasible project alternatives and anticipated losses in resource productivity. It is a document written to solicit public feedback; drafts as well as final copies must be made available to individuals or organizations at a cost not exceeding the charge of reproducing the materials. More important, in order to expand public access, agencies are encouraged to publicize the existence of the documents by placing notices in local newspapers or contacting organizations with recognized environmental interests. Legislation has also obliged other agencies to facilitate public scrutiny. For example, The Energy Re-organization Act of 1974 mandated the collection, preparation and dissemination of information on energy alternatives "to provide the free exchange of ideas and criticisms that is essential to. . . public understanding" (USCA, 1976). Provisions for access to documents open possibilities for citizens' intervention and give environmentalists a legal instrument to influence or delay projects. Such provisions, however, do not in themselves assure participation.

On the working level, these procedures are often limited. They are passive, leaving the burden on individual citizens to inform themselves. Even when agencies set up procedures to facilitate public access, the costs of obtaining information are high: keeping up-to-date on notices in the Federal Register is time-consuming and obtaining information may include transcript or duplication costs. There exist few provisions for staff assistance to help those who do not know exactly what documents they need.[1] Thus, while open information about impending decisions is crucial, demands for participation extend beyond simple access to documents. In response, a number of legislative and administrative procedures provide opportunities for interested

[1]Most requests on the Freedom of Information Act must be explicitly stated to be covered by the Act's provisions (10 C.F.R. § 9.8. 709.8).

groups to inform agency deliberations about their concerns, values and preferences.

Participation in Rule-making and Adjudicatory Procedures

Legislation delivers priorities and broad mandates to administrative agencies, but rule-making procedures determine the technical details and specific choices actually shaping outcomes. Thus, the public must participate in these procedures if it is to have any influence. Rule-making includes requirements for public notice and comment through the Federal Register. Public interest groups use this process to monitor new rules and standards and to encourage revisions. However, notice-and-comment procedures, while frequently used, tend to favour established groups who are already sufficiently well-informed to recognize the bearing of specific actions on their interests. Moreover, agencies are not bound by public comments and can ignore them. Agencies are under continued pressure to adopt a more activist role in seeking public input and the most frequent response is to expand the public-hearing process. Public hearings may have several functions (Heberlein, 1976). Often they simply present an opportunity to provide interested parties with information about proposed projects; reports, slides and graphs are common accoutrements. Agencies also use public hearings to defuse anticipated conflict, hoping to avoid litigation. When decisions are non-controversial or very narrow in scope, hearings can degenerate into administrative rituals. However, at times public hearings are genuinely interactive—a means to discover the ideas of participants as a basis for agency action.

More formal than public hearings are the adjudicative procedures required for the granting of licences and franchises in such activities as offshore oil-drilling and power-plant siting. Although citizens' group have appeared as witnesses and presented *amicus curiae* briefs, adjudications normally involve only the limited participation of those parties directly affected. Characteristically, lawyers argue the specific claims of utility companies and displaced property owners, while citizens concerned with the less specific implications of the impending policy are excluded. However, when citizens do participate in adjudications, their influence may actually be greater than in the less rigid setting of public hearings (Boasberg, 1977, pp. 132-134; Cramton, 1977). Thus, an important and unresolved issue in recent discussion about adjudicative procedures is the question of standing: Who may be included in these proceedings? Efforts to resolve this question and to expand the use of public hearings and adjudications as channels for participation are evident in changes in the institution of 'public intervenor'. The right of citizens' representatives to intervene as 'full parties' in adjudicative hearings has been clarified and extended by recent legislation; however, the ability to intervene is significantly restricted by the right of agencies to determine who may participate and what issues may be contested. For example, when intervenors petition to participate in a power-plant licensing hearing, they must state their interest in the proceeding, estimate the expected effect of the proposed action on their interest and identify these aspects of the conflict in which they intend to intervene. The Nuclear Regulatory Commission retains discretionary power to limit the presentation. Moreover, agencies usually rely on the willingness of a group to bear the costs of intervention as one measure of their commitment. Yet intervenors often lack the necessary financial and technical resources to present a convincing argument. Businesses consider intervention in regulatory procedures to be a necessary investment and can take tax deductions or add residual costs to the price of their products (Goetz and Brady, 1975). Concerned citizens, however, quickly discover the high cost of acting on their beliefs, especially in technical cases, where the purchase of scientific expertise compounds routine costs of time and travel. Given this situation, Senator Edward Kennedy has argued that public financing of citizens' intervention is necessary to ensure the right to participate (Kennedy, 1975).

Several groups have broached the idea of financing intervention. The National Academy of Sciences (NAS) Committee on Environmental Decision-making proposed that the Environmental Protection Agency (EPA) provide financial support to groups or individuals who can contribute to rule-making in adjudicatory proceedings (NAS, 1977). The Toxic Substances Control Act states that the administrator may provide compensation for expert-witness fees, as well as the costs of participating in rule-making proceedings (USC, 1976 a). The National Science Foundation's 'Science for Citizens' Programme proposed funding experts to work with citizen's groups (USC, 1976 b).

In order to consolidate various efforts to fund intervenors, Senator Edward Kennedy, in 1976, introduced the Public Participation in Government Act.[2] This Bill (which was not passed) would have authorized federal agencies to reimburse citizens' groups for the costs of participating in administrative proceedings. The controversy surrounding this bill reflects the general ambivalence about federal efforts to encourage expanded public influence. Proponents of the bill felt that greater participation would increase public accountability, encourage agencies to be more thorough and educate citizens about the process of decision-making. Arguments against the bill revolved around both practical and ideological points. Opponents warned of the high costs of such a programme and noted the difficulties of distinguishing between deserving and undeserving groups. Liberal interpretation of the law, they argued, would open agency doors to narrowly-based interest and to a cacophony of conflicting demands. It might produce not simply greater

[2] See testimonies of Senator Edward Kennedy, James Delong, David Marlin and Joseph Onek, in *Public Participation in Government Proceedings Act of 1976, Hearings* (USC, 1976c).

information, but the administrative nightmare of publicly financed opponents obstructing federally supported projects.

Social Research as a Means to Inform Decision-makers

Participants in hearings and adjudications are often criticised for not representing the public or being more concerned with their own special interests. Environmental groups, for example, are accused of elitist prejudices that neglect the employment and economic consequences of environmental policies. The Government and its agencies, it is argued, may be more representative of the public interest than those citizens' groups that have the power to effectively use available participatory channels (Clark and Stankey, 1976). This view has stimulated experiments, based on techniques of social research, that are directed towards providing agencies with information about actual public preferences.

Most of these techniques involve opinion polls. But in 1970, the US Forest Service employed a technique called 'codeinvolve' to uncover attitudes about its proposed use of DDT to protect Douglas Firs from moths (Campbell and Stanley, 1963).'Codeinvolve' is an applied content analysis system designed to transfer large quantities of information from written statements in diverse sources into an easily summarized form for policy review. After identifying questions of significance to the decision-makers, researchers then gathered and coded all relevant statements (from editorials, petitions, letters, etc.) on the issue. Condensing these scattered statements into clusters of opinion, the system offers a wider spectrum of attitudes than may emerge from other procedures, such as public hearings. Moreover, by counting the frequency of a given view as it appears in written statements, the decision-maker can get a measure of its prevalence. The Forest Service used 'codeinvolve' data to support its request to the Council on Environmental Quality (CEQ) for permission to use DDT. The study found considerable support for the use of the spray because people were concerned about the economic implications of moth damage and perceived very limited alternatives. Nevertheless, the CEQ rejected the Forest Services' proposal, giving greater weight to the concerns of environmentalists who were strongly opposed to DDT.

In another experiment, a group of scholars of the Environmental Science and Engineering Programme at UCLA experimented with a questionnaire and series of workshops with community interest groups. These groups were provided with opportunities to indicate their uncompromised preferences in choosing energy alternatives. Following the workshops, they were asked what trade-offs they were prepared to make after hearing more about the technologies and the concerns of their interest groups. This is one of a number of efforts to develop 'environmental mediation' techniques. Modelled on labour-management negotiations, this form of mediation requires a neutral 'third party' concerned with structuring an environment in which parties can negotiate. This presents a compelling alternative to the costly process of litigation. Moreover, its proponents claim environmental mediation is significantly more successful than public hearings in promoting constructive dialogue and in reducing conflict. Indeed, mediation procedures have played a significant role in resolving a number of environmental disputes (EIA, 1978). However, they require that both parties have sufficiently strong incentives to resolve their differences and that they are prepared to make reasonable compromises.

Another experiment, following from the confusing role of scientific conflict in environmental disputes, has been the proposed 'Science Court', a quasi-judicial procedure in which scientists with different views, on issues such as nuclear safety or the effect of high-voltage transmission lines, would argue before 'impartial scientific judges' (Presidential Group on Anticipated Advances in Science and Technology, 1976). In this proposed forum, debate would be limited to facts, leaving social-value questions to the political arena. But the opinion of the Science Court is expected to have enough authority to provide the basis for informed policy decisions. The democratic control of technology, its proponents claim, follows from establishing "the truth among conflicting claims of scientists". It assumes that scientists, through adversary procedures, can reach agreement and that this will lead to public consensus.

Although social research techniques may help to clarify public sentiment, most are participatory only in the most limited and indirect sense. Social research is, however, increasingly used as a way to inform decision-makers of citizens' preferences.

Public Advocacy Institutions

Several agencies have interpreted legislative mandates for increased public participation by forming public councils and citizen advocacy institutions. Offices of public council are intended as catalysts to bring together concerned citizens and agency decision-makers (Boasberg, 1977, pp. 73-76). Consumer advocacy institutions within agencies have proliferated. In California, the Office of Public Advisor to the Energy Resources, Conservation and Development Commission ensures that adequate notice-and-comment procedures are followed prior to decision-making. The success of these efforts has prompted legislative proposals to create an Agency for Consumer Advocacy, independent of other federal agencies, that would have no regulatory responsibilities. This, if it can overcome the doubts of many Congressmen, has the support of most public interest lobbies and the endorsement of the President.

Citizen advocacy institutions were the model for a plan developed by the National Science Foundations (NSF), the 'Science for Citizens' programme. This was intended to improve the public's understanding of controversial technical issues, to encourage the involvement of scientists in public policy and to facilitate through federal support the access of citizen's groups to scientific expertise (Culliton, 1976; Primack, 1976). This last goal, essentially the technical equivalent of public council, proved highly controversial. To its proponents, providing expertise to citizens' groups is necessary to ensure a balanced discussion of controversial technical questions. The federal government and private corporations have ready access to costly technical advice, while citizens' groups must rely on scientists willing to volunteer their expertise. Expertise is a political resource. NSF-funding of citizen-group consultants could reduce the differences between agency and citizen authority in technical controversies. Critics, however, feared the potential for greater litigation, conflicts and delay if scientists became 'paid advocates' for citizens' causes.

Most participatory measures are directed to informing the public, or to informing the decision-makers, about preferences. They pose little real challenge to existing decision-making authority. A third set of procedures has possibilities for negotiation and compromise and is intended to provide more substantial and collaborative forms of public participation.

Collaborative Participation

A thin line exists between informative and collaborative participation. Concerned citizens have contributed formative ideas to the decision-making process and those more disgruntled have used available information to block proposals through protest or litigation. However, in the formal participatory procedures discussed so far, government agencies retain the right and the power to disregard citizens' concerns. What distinguishes more collaborative modes of participation is the inclusion of public representatives, not simply as informants but as partners in negotiations with some power to assure that agencies act on the basis of their preferences. Collaborative procedures include the opportunity for participants to evaluate the information presented and to negotiate the resolution of conflicting ideas and priorities. Perhaps the most striking feature of such procedures, however, is their rarity. Provisions for collaboration within agencies are found primarily in the use of advisory boards and commissions. However, the most powerful means of direct public influence has been through citizen litigation and referenda—channels of participation that directly challenge administrative procedures.

Advisory Boards

Advisory boards are the most frequent form of collaboration. In 1975, forty-five government agencies had 1,267 advisory boards with a total of 22,256 members. Agencies are not formally obliged to implement the recommendations of these Boards but consistent arbitrary action is rare, partly because Board membership is often representative of powerful groups whose support may be necessary for effective policy implementation. A survey of federal agency advisory boards found that nearly 50 per cent of the board members were representatives of industry while consumer and environmental groups comprised less than 7 per cent (Sullivan, 1976). Agencies justified this practice by claiming that citizens' groups were disinterested, lacking in technical qualifications or simply inappropriate. Many citizens' groups, however, felt that agency policy concerning membership on boards was a way to avoid outside pressure.

Citizen Litigation

The most effective countervailing force to administrative autonomy in the United States has been citizen litigation—a major channel of public involvement, especially in environmental policy. When dissatisfied with their influence through hearings or other administrative channels, citizens force their opinions through the courts. They have directly sued both government agencies for failure to regulate technologies that damage the environment, and private industries for developing technologies in violation of environmental standards.

'Standing' to sue varies from state to state but the rules have been liberalized. In the past, to bring a case to court, an individual or group had to demonstrate that either their property or legal rights had been 'invaded'. Direct injury was a prerequisite to court involvement, in effect restricting citizens' suits to cases of product liability and emminent domain. Yet technological injuries are often vague and indirect; pollution, for example, may have subtle and cumulative effects, widely suffered by an indeterminable number of individuals. In recognition of this, the restrictive doctrine of 'standing' has been modified and liberalized to allow a private citizen, without an alleged personal economic injury, to present a grievance as an advocate of the public interest. Moreover, environmental litigation has brought pressures to include as cause for standing not only damage to property rights but some of the more subtle health and aesthetic impacts of technological progress (Di Mento, 1977).

The National Environmental Protection Act and the Environmental Impact Statements' (EIS) requirement opened the door to considerable litigation. Between 1970 and 1975, there were 332 suits against federal agencies for inadequate EIS reports. Only 64 were sustained but the threat of litigation has remained a powerful participatory tool.

The Referendum

The referendum is a vestige of direct participatory democracy within representative governments, an instance of collaborative participation that may effectively pre-empt administrative authority. Referenda are a prominent and increasing feature of the United States political landscape. There are thousands of referenda each year, ranging from local issues such as property taxes to environmental concerns such as airport expansion or nuclear power plant development (Shepard, 1975). Twenty-one states have procedures for placing issues on the ballot by obtaining a given number of signatures; twenty-three states, as well as hundreds of cities, permit direct legislation by voting; and 38 states have some form of referenda permitting voters to review laws passed by the legislature. These procedures encourage citizens to initiate legislation or to repeal existing laws that violate the public will.

A most striking use of the referendum occurred in 1976, when critics placed nuclear control initiative proposals on ballot in eight states: California, Oregon, Arizona, Colorado, Ohio, Missouri, Washington and Montana. In June 1976, Californian citizens voted on a proposal which directly challenged administrative decision-making on nuclear safety. If it had passed, Proposition 15 would have required two-thirds of the California legislature to approve both the effectiveness of nuclear power plant safety systems and the adequacy of waste-storage facilities, prior to the development of a nuclear programme. This use of the direct-citizen vote was essentially an effort to return technical decision-making to a more participatory framework, by allowing a state legislature to control matters usually delegated to administrative agencies. Proposition 15 was, however, defeated, as were similar propositions in the other states. Although polls in California and Oregon indicated that a slight majority of 53 per cent favoured restrictions on nuclear development, the public proved more vulnerable to the arguments from utilities, and their associated industries, about the consequences of the nuclear programme for economic progress and employment.

THE LIMITS OF PARTICIPATORY REFORMS: PROCEDURAL BIASES

Administration agencies implement participatory mechanisms with an ambivalence that is reflected in procedural biases. These often limit the effective implementation of the participatory objectives that are mandated in recent legislation. And they suggest how agencies, while responding to political demands for greater public involvement, seek to reinforce the centralized control that they believe necessary for administrative efficiency.

The Problem of Vagueness: Who Should Participate?

One consistent feature of the Congressional legislation requiring participatory procedures is their lack of specificity. Most legislation comments on the general circumstances under which participation shall occur but avoids more tangible questions of who should participate and the extent of outside influence. Implementation of participatory measures immediately raises the problem of who will be involved, yet none of the participatory efforts reviewed have comfortably dealt with this. Thus, representation may be biased towards small, but intense, well-organized groups with few claims to represent 'the public'.

Some regulations are more explicit than others in defining 'the public'; in the Nuclear Regulatory Commission's Energy Facility Licensing procedure, an intervenor is "one whose interest may be affected by the proceeding". Those not qualified to become official intervenors may make "limited appearances" at hearings "at the discretion of the Board". Through these rules, the Commission believes that "all members of the public are assured of the right to participate by a method appropriate to their interest in the matter". But since the regulations fail to provide clear guidelines for determining appropriate interest, the potential for manipulation remains.

Administrative discretion also extends to critical procedural decisions. It is the agencies that select members of advisory committees; the people who may speak at open committee or staff meetings; the individuals to contact when legislation calls for notification of the 'people directly affected' by an action; and the individuals who have administrative standing to initiate grievance procedures.

Lack of procedural specificity allows manipulation and creates an unstable basis for public participation. If agencies do not specify procedures embodying the Congressional intent, efforts to increase public involvement can be effectively undermined by an uncertain, discretionary and arbitrary environment.

The Problem of Reactive Participation

Timing of public involvement in the decision process further limits the potential influence of the public. Most opportunities for participation are offered late in the decision-making process; usually citizens can only react to proposals already developed and accepted by agency staff.

When administrators solicit public opinion on reports, they do do after these documents have been compiled (with the exception of Environmental Impact Statements). Furthermore, public expression is permitted only just prior to a proposed change of rule, that is, only after agency staff and applicants have arrived at a preliminary consensus. Because public involvement occurs so late in the process, the burden falls on public intervenors to prove the potential problems of a project (even though it is the applicant who formally bears the 'burden of proof'). If the public enters the decision-making process after significant choices are already made, its only input, other than acquiescence, is the veto. Consequently, the public becomes a 'nay sayer', perpetuating prejudice on both sides: government is the 'enemy'; the public only plays a negative role.

The Problem of Cost

Finally, the problem of cost indicates further bias against effective implementation of participatory procedures. Participatory measures such as intervention have considerable potential for influencing decisions but to effect a decision, an intervenor must be able to present a polished, informed and technically strong argument. Legal and scientific expertise is necessary to prepare and to effectively argue a case. According to one estimate, the legal and technical costs of a full-scale citizens' intervention in a rule-making hearing may be US$75,000 to $100,000.[3] If public involvement is limited to only the most organized and wealthy interest groups, this may effectively subvert the 'right' of participation.

CONCLUSION

Deeply-rooted democratic values hold that the public has a 'political right' to information and that more direct citizens' involvement will reduce conflict, enhance trust in agency decisions and improve the quality of decision-making. Although admittedly time-consuming, it is believed that public input into decisions will lead to substantive contributions to environmental protection by raising new issues and serving to counterbalance narrow agency biases. Active participation, it is argued, is necessary to assure that policies reflect public preferences; and procedural justice is necessary to foster eventual public acceptance of government decisions.

However, those with administrative responsibility are brought up to place high priority on efficiency. They resent interference and conflict as an obstacle to their efficient implementation of policies they perceive to be in the public interest. Participatory measures, it is feared, will only encourage demands for primitive interests and self-appointed public representatives who lack the technical knowledge to assess decisions. They can seriously impede the decision-making process, by causing costly delays and short-sighted negative decisions. Unfettered expertise, in this view, is central to assuring the efficiency of administrative procedures (table 1).

These are the arguments in the participatory debate. The actual experience concerning the effect of greater participation hardly resolves the dilemma. In some cases, greater participation appears to improve public attitudes towards agencies, while failing to reduce conflict. For example, in 1970, the Corps of Engineers developed an experiment to expand participation

[3]Ernest Gellhorn, in *Public Participation in Administrative Proceedings*, states: "Estimates of the fees charged by attorneys in 'typical' cases began at US$4,000 for comparatively simple ICC tariff proceedings. Fees for formal rule-making proceedings at the FDA are estimated, conservatively, at $4,000. Intervention in a major FTC proceeding might cost $100,000 or more in fees. Major rate proceedings in the CAB, FCC or FPC and major licensing contests in the FCC or FPC often generate fees in excess of $100,000." (Gellhorn, 1972).

Table 1: Public participation

Mechanisms	Justifications	Critiques
To improve public access to information.	The public has a right to know. Information will increase confidence in agency decisions and reduce conflict.	Too great information will generate greater public concern, increasing conflict and reducing efficiency.
To channel information from public to policy-makers.	Information from the public to improve the quality of decision-making. It may make substantive contributions to public safety and environmental protection, raising new issues, providing new assessments and adding useful data. Public input can also serve as a check on administrative discretion, balancing narrow agency biases, thereby improving procedures.	Opportunities for involvement will cause costly delays that will only raise consumer costs. It will contribute to the inundation of material from illegitimate, non-representative interests.
To foster collaborative participation.	There is no monolithic public interest that can be represented by an agency. Thus, given the non-technical nature of the issues, participatory decision-making is a 'right' in a democracy. It will also reduce conflict, promoting acceptance of government decisions.	Opportunities for full participation will provoke endless demands by multiple, non-representative private interests with personal biases or by 'self-appointed' environmentalists. This may lead to short-sighted policy, favouring the *status quo*.

in the hopes of reducing the continued controversy that surrounds its massive projects, with their inevitable effect on natural resources. Many of the Corps' projects fall under the Federal Water Pollution Control Act, which requires agencies to document their compliance with legislative mandates for public participation and gives citizens the right to petition for public hearings if none are planned (EPA, 1973; Nelkin, 1977). Previously the Corps had used hearings mainly to inform the public of well-formulated plans. But in the early 1970s, they announced a new 'open' policy to synchronize participation through public hearings with the evolution of particular project proposals (Dodge, 1973; Hanchey, 1975). Hearings were first held at the exploratory stage of planning in order to obtain information on the scope of issues needing consideration. Then, a second set of hearings were called to evaluate public support for various options. Prior to implementation, a third series of hearings were intended to ensure public support by allowing last-minute modifications. The Corps also held meetings with small groups of community leaders and large groups of concerned citizens, hoping to contact the full range of interested parties. Even critical environmental groups lauded these procedures.

Despite apparent success, however, the Corps soon retreated from its aggressive pursuit of citizens' involvement. Although the hearing had improved its public image and provided useful dialogue with various interested parties, the public involvement experiment did not, as anticipated, prevent conflict; the conflicts reflected not simply misunderstanding, which might have been corrected through dialogue, but rather basic disagreement over the relative values of technological development and environmental preservation. Participatory mechanisms further exposed, and thus intensified, those disagreements (Mazmanian, 1976).

There are certainly many cases where public participation has caused significant delay or obstruction of projects; but in other cases lay participants have provided support. The Cambridge Review Board (a lay committe formed to assess the environmental and health risks of a proposed, recombinant DNA facility) unanimously supported continuation of the research. Obstruction may be more serious problem, as a result of controversy rather than structured participatory arrangements. And inevitably, while greater participation is a source of costly delay, there are many examples where improved environmental quality and safety have resulted directly from public demands.

At present, administrative priorities of efficiency and autonomy cast doubts on any real commitment to expanded participation. Participatory mechanisms are directed more towards co-opting public support than towards changing the balance of existing authority, more towards creating informed consent than towards expanding democratic choice. Often, participatory procedures divert action into paperwork—hardly a reasonable substitute for genuine involvement.

However, the participatory impulse continues to place demands on administrative institutions in Western democracies. New mechanisms to broaden public involvement and to bring new groups into the decision-making process have proliferated. In Europe, efforts to foster greater public acceptance of science and technology through expanded public involvement include the Windscale public inquiry in England, the extension of the Study Circle concept to consider energy questions in Sweden, public debates on nuclear power in Austria, reforms of local public inquiry procedures in France, and the early publication and dissemination of 'policy intentions' in Holland.

As governments grope for ways to meet demands for greater public involvement, many difficult questions remain. What form can lay participation take in complex technical areas? Who should be involved? What will be the impact of this greater public role on technological development? On administrative efficiency? Attitudes towards participation remain ambivalent. As a concept, participation is a source of legitimacy but, as a procedure, it may be inefficient and obstructive. There remains considerable fear that greater public involvement will virtually paralyze technological progress and economic growth. Nevertheless, the implementation of policies for science and technology, and the legitimacy of the political authorities responsible for them, may ultimately depend on the ability to find satisfactory mechanisms for greater public participation and control. Indeed, if a World's Fair in the 1970s were to focus on science and technology, it would, perhaps, adopt the theme: "Science Finds—Industry Applies—Man Controls."

References

Boasberg, T. (1977). *Implications of NSF Assistance to Non-profit Citizen Organizations.* Washington, D.C., pp. 73-76 and 132-134.

Campbell, D. and Stanley, J. (1963). *Experimental and Quasi-Experimental Designs or Research.* Rand McNally College Publishing Co., Chicago.

CEQ/Council on Environmental Quality (1973). *Preparation of Environmental Impact Statements: Guidelines.* Federal Register, Vol. 38 No. 147, 1 August 1973, pp. 20550-20556.

Clark, R. and Stankey, G. (1976). *Analyzing Public Input to Resource Decisions: Criteria, Principles and Case Examples of the Codeinvolve System.* Natural Resources Journal, Vol. 16, January 1976, pp. 213-222.

Cook, T. and Morgan, P. (Eds.) (1971). *Participatory Democracy.* Canfield Press, San Francisco.

Cramton, R. (1977). *The Why, Where and How of Broadening Public Participation in the Administrative Process.* Georgetown Law Journal, Vol. No. 3, February 1977, p. 529.

Culliton, B. (1976). *NSF, trying to cope with Congressional Pressures for Public Participation.* Science, 191, January 1976, p. 274.

Di Mento, J. (1977). *Citizen Environmental Litigation and Administrative Process.* Duke Law Journal, 22, pp. 409-452.

Dodge, B. H. (1973). *Achieving Public Involvement in the Corps of Engineers Water Resources Planning.* Water Spectrum Bulletin, Vol. 9, No. 3 June 1973, p. 449.

EIA/Environmental Impact Analysis Review, 2, October 1978. The MIT Press, Cambridge, Mass.

EPA/Environmental Protection Agency (1973). *Public Participation in Water Pollution Control.* Federal Register, Vol. 38, No. 163, 23 August 1973, Part III.

EAS/Federation of American Scientist (1975). Professional Bulletin, Vol. 2, No. 5, p. 5.

Galbraith, J.K. (1967). *The New Industrial State.* Signet Books, New York.

Gellhorn, E. (1972). *Public Participation in Administrative Proceedings.* The Yale Law Journal, Vol. 81, No. 3, January 1972, p. 367.

Goetz, C. and Brady, G. (1975). *Environmental Policy Formation and the Tax Treatment of Citizen Interest Groups.* Law and Contemporary Problems, Vol. 39, No. 4, pp. 211-231.

Hanchey, J. (1975). *Public Involvement in the Corps of Engineera Planning Process.* National Technical Information Service, US Department of Commerce, Springfield, VA.

Heberlein, T. (1976). *Some Observations of Alternative Mechanisms for Public Involvement: The Hearing, Public Opinion Poll, the Work Shop and the Quasi-experiment.* Natural Resources Journal, 16, January 1976, pp. 197-212.

Keim, D. (1975). Participation in Contemporary Democratic Theories. In *Participation in Politics,* J. R. Pennock and J.W. Chapman (Eds.), Leiber Atherton, New York, p.l.

Kennedy, Sen. E. (1975). Statement in Congressional Record, 121. 174, 20 November 1975.

Kolata, G. (1975). *Freedom of Information Act: Problems at the FDA.* Science, Vol. 189, July 1975, ppp. 32-33.

Marris, R. and Rein, M. (1967). *Dilemmal of Social Reform: Poverty and Community Action in the US.* Atherton Press, New York.

Marris, R. (1964). The Economic Theory of Managerial Capitalism. Free Press, New York.

Mazmanian, D. (1976). Participatory Democracy in a Federal Agency. *In Water Pollution and Public Involvement,* H. Doeksen and J. Pierce (eds.).

Mazur, A. (1973). *Disputes Between Experts.* Minerva, XI, pp. 213-262.

NAS/National Academy of Sciences (1977). *Decision-making in the Environmental Decision-making Agency II.* NRC Committee on Environmental Decision-making, Washington, D.C., pp. 102-113.

Nelkin, D. (1971). *Nuclear Power and its Critics.* Cornell University Press, Ithaca, New York.

Nelkin, D. (1975a). *Jetport.* Transaction Books, New Brunswick, N.J.

Nelkin, D. (1975b). *The Political Impact of Technical Expertise.* Social Studies of Science, February 1975.

Nelkin, D. (1977). *Technological Decision and Democracy.* SAGE Publications, Beverley Hills, California.

Nelson, R. (1977). *The Moon and the Ghetto.* W.W. Norton, New York.

Presidential Group on Anticipated Advances in Science and Technology, Task Force of (1976). *The Science Court Experiment.* Science, 193, August 1976. p. 653.

Primack, J. (1976). Testimony. US Congress, Senate, Special Subcommittee. In *NSF, National Science Foundation Authorization Legislation.* G.P.O., Washington, D.C., 3 March 1976.

Schilling, T. (1974). On the Ecology of Micromotives. In *The Corporate Society,* Robin Marris (Ed.) John Wiley and Sons, Inc., New York, p. 63.

Shepard, W.B. (1975). *Participation in Local Policy-making: The Case of Referendum.* Social Science Quarterly, 56, June 1975, pp. 55-70.

Spiegel, H. (Ed.) (1968). *Citizen Participation in Urban Development.* National Institute for Applied Behavioral Sciences, Washington, D.C.

Sullivan, J. (1976). *Public Participation in Federal Energy Advisory Committees.* Draft Report to the National Council for the Public Assessment of Technology. (Mimeo.)

USC (1946). *Administrative Procedures Act.* 5 USC, ξ 551.

USC (1975). *Freedom of Information Act.* 5 USC, ξ 552, Supp. II.

USC (1976a). *Toxic Substances Control Act of 1976.* 15 USC, ξ 2619.

USC (1976b). *National Science Foundation Authorization Act of 1977.* ξ 5, 42 USC, ξ 1862. Amending 42 USC, ξ 1861 (1950).

USC (1976c). US Congress, Senate, Committee on Government Operation. In *Public Participation in Government Proceedings Act of 1976, Hearings,* G.P.O., Washington, D.C., p. 45.

USCA (1976). ξ 5817 (e).

Wade, N. (1972). *Freedom of Information Officials Thwart Public Right to Know.* Science, Vol. 175, February 1972, pp. 498-502.

Yin, R. *et al.* (1973). *Citizen Organization: Increasing Client Control over Services.* Rand Corporation, Santa Monica.

CHAPTER 6

Integrated Physical Planning in Sweden

Per Wramner

INTRODUCTION

THIS chapter deals with the environmental aspects of integrated physical planning in Sweden, where integration is both horizontal (between different interests and sectors in society) and vertical (between national, regional and local interests and tasks). Physical planning in Sweden is, by tradition, principally a municipal responsibility. During the past decade, however, the State has come to play a more active part in the process. State planning covers, amongst other things, infrastructure, management of natural resources and conservation of the environment. State planning also covers the economic sectors, employment, etc., which are of fundamental importance to physical planning. Various methods are used to integrate different elements of the planning process. Although this integration has developed gradually and improved during recent years, it has not yet achieved its definitive form.

In Sweden, the best conditions for the close co-ordination of different planning activities exist at municipal level. The municipalities have extensive powers in physical planning, as well as exercising firm control over the implementation of plans, particularly in urban areas. This control is guaranteed by an active land policy and also by the powerful influence they exert on the planning of housing supply and investments in streets, roads and amenities. Thus, the planning and executive instruments available to the municipalities, and the proximity of the planners to those who are affected by decisions, combine to present good opportunities for well-balanced social development under the aegis of municipal authorities.

ADMINISTRATIVE FRAMEWORK AND THE MAIN ELEMENTS OF THE SWEDISH PLANNING SYSTEM

At the national level, planning is carried out by several different ministries and central agencies. All decisions at the governmental level are formally taken by the Government as a whole. Major decisions are invariably preceded by exhaustive consultations between the ministries concerned. These consultations are made easier by the fact that the ministries are small units whose main responsibility is the definition of policy guidelines, while most of the responsibility for the implementation of policy rests with the central agencies. The Ministry of Housing and Physical Planning is responsible for physical planning at central government level, while the National Board of Physical Planning and Building is responsible for implementation. The corresponding responsibility in the field of environmental protection rests with the Ministry of Agriculture and Environment, and the National Environment Protection Board.

The twenty-four State County Boards are the administrative agencies of the Government at the regional level. These Boards act in a supervisory and regulatory capacity, hearing appeals and ratifying plans taken by the municipalities. Another important task is to co-ordinate State and municipal planning within the county. The State County Boards have recently been re-organized with a view to the better co-ordination of, among other things, the different branches of planning. Each Board has separate departments of physical planning and environmental protection, each with specialist staffs. The role of the State County Boards as co-ordinating bodies is accentuated by

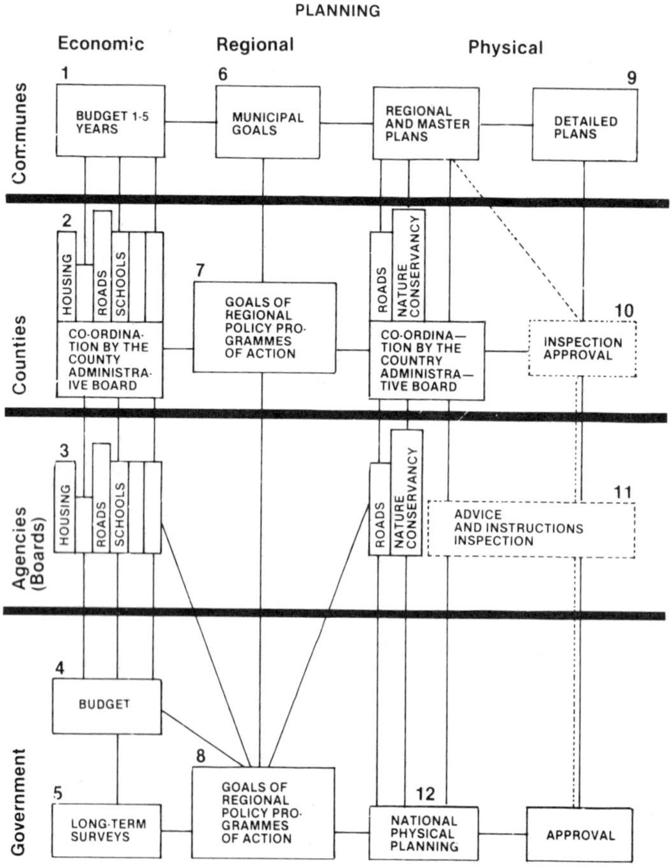

Figure 1: The Swedish Planning System

Planning normally conducted on a sector-basis is represented by vertical columns (vertical planning), while planning involving more than one sector (cross-sectoral or horizontal planning) is indicated by horizontal columns.

Economic Planning

Municipal level (1) Municipal economic planning is both sectoral and cross-sectoral (in the annual budget and municipal long-term economic planning).

County level (2) Hitherto, most economic planning at county level has been on a sectoral basis. The most important decision-making powers are vested in the State County Boards. This planning governs a large proportion of investments through the medium of outline investment plans, quotas, etc.

Agency level (3) Planning is conducted on a sectoral basis. Outline investment plans and quotas are distributed to the counties for further distributing to the municipalities. In certain cases the outline investment plans are conveyed directly to the municipalities.

Government level. Planning is mainly cross-sectoral and is primarily manifested in the annual budget (4). A balance is struck between different sectors. Long-term surveys (5) provide a foundation for long-term assessments.

Physical Planning

Municipal level. (9) Physical planning at municipal level is cross-sectoral. Both detailed plans (building plan, town plan) and outline plans (master plan and regional plan) co-ordinate the interests of different sectors in the utilization of land.

County level (10) At county level, physical planning in the formal sense of the term is mainly concerned with roads and nature conservancy. Otherwise, State involvement at this level takes the form of inventories, advice to municipalities and inspection and approval of certain municipal plans.

Agency level (11) Physical planning, in the formal sense of the term, is mainly conducted in the road and nature conservancy sectors. To these are added the inspection of municipal plans and the provision of forms and other basic material for planning at subordinate levels.

Government level (12) Cross-sectoral physical planning is conducted in the form of national physical planning. Guidelines for land and water management are drawn up by the Government and the Riksdag. Certain municipal plans are subject to Government ratification.

Regional Policy Planning

The main emphasis of regional policy planning is on the counties (7), the aim being to influence regional development, e.g. by a co-ordination of sectoral planning in accordance with the goals laid for regional policy. The communes influence these goals through their pronouncements, etc. (6), and the goals are then evaluated politically by the Government and the Riksdag (8).

Source: Planning Sweden (1973).

the fact that they are supervised by a board of laymen, with the Provincial Governor as Chairman.

At local level, there are almost 280 autonomous municipalities with powers of taxation and independent decision-making in important sectors of public policy. This high degree of local autonomy is one of the outstanding features of national administration in Sweden. Land-use planning and the detailed control of new construction are primarily the responsibility of the local authorities. Within the municipalities, the Executive Committee has the main responsibility for planning. Its co-ordinating function is reflected by its responsibility for general physical planning, long-term economic planning and the drafting of the annual municipal Budget. The staff of the municipalities generally includes experts in physical planning but not in environmental protection.

The administrative framework of the Swedish planning system is summarized in figure 1, illustrating in schematic form the mutual relationship of the various forms of planning. The Swedish planning system is not totally integrated but there is a clear functional connection between its different elements. The forecasts with which regional policy planning begins are to a large extent based on the economic, long-term surveys. The inventories of requirements on which national physical planning is based are founded mainly on the same assumptions and assessments as those of the long-term surveys. But it is worth noting that time horizons differ; whereas the long-term surveys cover a planning period of five years, regional policy planning and national physical planning look 15 and 30 years ahead, respectively.

Thus, to a degree, the forecasting data of the long-term surveys at present provide a general starting point for other kinds of community planning. But it is likely that this situation will eventually be partly reversed, so that both regional policy planning and national physical planning (after they have been further developed and their aims defined more exactly) will figure as permanent elements of the long-term surveys.

Regional policy planning is chiefly concerned with the development of built-up areas. The goals of regional policy, adopted by the Riksdag (the Swedish Parliament), are to be seen as a political instrument, a social declaration of intent. They are expressed in planning outlines for the future development of the counties, stated in terms of population, and in a plan for the development of the regional structure, which indicates the hierarchy of urban centres. In drawing-up these plans, consideration is given to environmental conditions in different communities. The goals of regional policy are to be regarded as a basis for physical planning. During recent years, regional transport planning and energy planning have become substantial elements in the planning system. In particular, the demand for energy conservation and the reservation of natural resources for energy production are of fundamental importance in integrated physical planning.

National physical planning has also become an important part of the Swedish planning system during the 1970s. Its principal task is to indicate the disposition of land and water as physical resources outside the built-up areas, insofar as there is competition between rival interests for the same resources, for instance between conservation and exploitation. National physical planning is an important instrument in the implementation of natural resources management and environmental conservation policies in Sweden.

LEGISLATION

An important basic principle in Swedish planning and building legislation is that the local authorities are mainly responsible for drawing-up plans and granting building permits. But there is a need to co-ordinate the planning of different municipalities since there are vital national interests (for instance nature conservancy) to be considered. Therefore, the State is also involved in the planning process, mainly in a supervisory role, and in national physical planning.

The use of land and water in Sweden is regulated by several bodies of legislations. Physical planning is mainly based on the Building Act which contains different planning instruments, for instance regional plans, master plans and detailed development plans. Informal plans are also widely used, for instance municipal-wide plans that provide a form of outline development control for non-urban areas. The land-allocation plan shows, in broad outline, present and long-term land use. The municipal structure plan states the regulations concerning building, where and how building may take place and where plots can be subdivided; it also sets out the planning measures envisaged. The Nature Conservancy Act makes provisions for safeguarding the interest of nature conservation and outdoor recreation. The Environment Protection Act contains regulations for protection against water and air pollution, noise and other nuisances, arising from the utilisation of land or buildings. Instruments, such as pre-emption and expropriation, are also available to municipal land policy.

NATIONAL PHYSICAL PLANNING

Aim

The aim of national physical planning is to achieve good management of the country's land and water resources in a national perspective. There is no single, nation-wide plan regulating where development can or cannot take place or how land and water areas are to be utilised for different purposes. Instead, national physical planning is a method of determining which demands on land and water resources are of national interest, as well as providing a system of plans and land-use regulations ensuring that vital national interests are taken into consideration. National physical planning involves the interplay of local authority planning and government guidelines. It is a continuous ongoing process that is affected by administrative and political decisions taken by local authorities, state boards and the government.

Background

Sweden is a sparsely populated country. For a long period, it was possible to satisfy the claims on land and water resources for different uses without inviting serious conflicts. Urban settlements occupy only 2 per cent of the land area. However, the claims on land for different uses, such as secondary dwellings (weekend cottages or summer bungalows), industrial expansion, open-air recreation, nature conservancy and the preservation of historical monuments, have increased substantially in recent years. Awareness of the scarcity of raw materials and natural resources, such as minerals and water power, has also grown. All this has touched off a great need to bring about a composite assessment of the country's assets and a planned system for the economic management of our land and water resources.

Proposals for physical planning at the national level have been presented on several occasions during the past few decades. A topic that came up in the mid-1960s was the location of large manufacturing plants outside existing industrial settlements. The proposals usually triggered off a lively debate. For example, there was the question of whether it was appropriate to site large oil refineries or nuclear power stations in attractive coastal areas used for bathing and recreation. Moreover, people became more aware of the fact that built-up areas of secondary dwellings were beginning to pose a threat to the most attractive recreational tracts along the coasts, in certain river valleys and in the mountain regions.

The municipalities had long been engaged in the physical planning of urban settlements, seeking to balance private and public interests and invoking various instruments of land policy for this purpose. But those instruments now stood out as inadequate within the extensive rural areas and demands were therefore raised for the community to plan outside the urban settlements.

The Conservation of Land and Water Resources

In the late 1960s, work commenced on investigating a system of physical planning at the national level. This work resulted in a governmental report in 1971 entitled *The conservation of land and water resources.* Different surveys were presented in the report, for instance proposed sites for polluting industries and areas of national interest for outdoor recreation, nature conservancy and the protection of historic buildings and ancient monuments. The report proposed, among other things, that certain coastal areas be exempted from the location of polluting industries and some completely wild areas in the mountains be protected from extensive development. Vacation-cottage development was to be restricted, particularly along the coasts of southern Sweden.

In 1972, the Government put forward a Bill, based on this report, with guidelines for the conservation of land and water resources, together with a countrywide programme of action for regional policy. In this way, the relationship between regional policy and physical planning was emphasized.

Decision by the Riksdag

The Riksdag gave its approval to the Government's proposals concerning the conservation of land and water resources and the way in which national planning work was to be carried out. Parliament's decision in 1972 emphasized an ecological approach as the basis for physical planning. An ecological perspective means that physical planning should assist society to develop within the limits set by our natural resources and that the environment and the diversity of natural ecosystems should be conserved to ensure rich possibilities of biological development in the future.

The most important part of the Riksdag's decision was certain general guidelines. These had a governing effect on municipal and state organs. They could be divided into two main categories concerning (1) certain activities, such as agriculture, forestry, outdoor life, tourism and nature conservancy and (2) certain geographically designated areas in Sweden. Furthermore, vacation-cottage development was to be planned in such a way that the natural resources available for recreation could be utilized by everyone. Much attention was given to areas of great value to nature conservation and outdoor recreation. Such areas of national interest should be dealt with carefully in local planning. According to the guidelines, particular attention was to be paid to agriculture and forestry. Where conflicts developed between agricultural interests and urban expansion, local authority planning should attempt to find solutions that entail as little development of high-quality agricultural land as possible.

A number of alterations were made to the legislation on physical planning, building and nature conservancy. The application for a building permit was made compulsory for practically all types of development. The suitability of new development in areas not covered by local development plans could, therefore, be subjected to scrutiny by the local authority. If development was not permitted in accordance with the regulations concerning the suitability of development in non-urban areas, then no compensation would be paid. Only when the existing land use was affected would compensation be paid. In accordance with the Riksdag's decision, the Building Act was provided with a supplementary provision (136a), stating that the Government had to approve the location of new industrial development of the type dealt with in national physical planning. The Government's approval can only be given where the local authority has given its prior approval (i.e. a local authority has the ultimate right of veto).

Work on the implementation of the guidelines, in accordance with the Riksdag's decision, has involved a progamming phase and a planning phase, in which local authorities have put forward their proposals to the State County Boards who have collated this material, added their own comments and presented it to the Government.

Municipal Planning

In 1973 and 1974, the municipalities wrote progammes for their planning work. These programmes were examined and approved by central administrative agencies, Government and the Riksdag in 1975. The municipalities then embarked on their planning and implementation work, mostly completed in 1977. This planning phase was concluded in 1979 with the decisions by Government and the Riksdag.

The planning work sought, first of all, to shed light on the development of building coverage in the whole municipality, mainly outside the urban settlements. Guidelines for such development were issued not only with future needs in mind for new roads, power lines, airports and other major civil engineering works, but also with allowances made for the interests of agriculture, nature conservancy and the preservation of historical monuments. These land needs were derived from material documented during the programme stage by the responsible central agencies, in co-operation with the State County Boards. These boards played a key role in the work by providing, for example, expert knowledge and basis material for use by the municipalities.

Secondly, in respect of those areas outside the urban settlements where the competition for land is keenest (in the coastal areas, river valleys, mountain ranges and around the larger lakes), the municipalities were required to draw upon the parliamentary guidelines to prepare comprehensive outline plans for land use, giving a picture of the current type of land use and the major changes the municipality intends to make or permit. Thus, the municipalities had to identify, on 1:50,000 topographic maps, the areas to be reserved in the long-term for agriculture and forestry and areas to be protected or preserved out of consideration for nature conservancy, the preservation of historical monuments or outdoor recreation. In addition, the municipalities had to stipulate, over a 10 to 15-year period, the alocation of large-sized manufacturing plants, roads, airports, harbours, artillery ranges, power lines or recreational facilities. About 10 per cent of the country's surface area has thus been subjected to this type of comprehensive, land-use planning.

Thirdly, the municipal planning work culminated in a report setting out the concrete implementation measures that need to be taken. Municipalities with jurisdiction over the country's most important farming districts, and over those primary recreational areas which the Riksdag singled out in 1975, were also required to present programmes to show how they intend to protect the land needs of agriculture and recreational life. During the programming and planning phases, the guidelines were adjusted and supplemented. The original aim and direction remained unchanged however. In many cases, the ecological aspects were emphasized. Stronger restrictions were imposed for instance on the use of agricultural land for development purposes, on the development of rivers for hydroelectric purposes and on vacation-cottage development. The common uniform planning work is now at an end but a number of important measures still need to be taken, such as the drawing-up of more detailed, land-use plans in certain conflict areas and the safeguarding of nature conservancy and outdoor recreation interests.

The Conservation of Land and Water Resources 2

Concurrent with the completion of the first stage of national physical planning, a second stage was introduced. A governmental report, *The conservation of land and water resources 2,* was published in 1979. This report has an even more pronounced ecological approach to the future use of land and water resources in Sweden. The report describes the land and water resources requirements of different activities. Proposals are also made concerning some supplementary and new guidelines, giving environmental aspects greater weight in the planning process, such as the expansion of continuous coastlines. Energy supply, mineral resources, water management, maritime environment—all are dealt with in greater detail than in the previous report.

Interest is shown in developing new areas for *agriculture and forestry,* primarily forestry plantations in marsh areas. These two sectors also exhibit an increasingly industrial character which seriously affects natural resources; examples are the intensive use of fertilizers, chemical pest control, plant breeding improvements and irrigation.

Areas for future research and ecological purposes are important for *nature conservancy.* Genebanks are required for flora and fauna in order to prevent genetic erosion. Interest in *outdoor recreation* (for example trekking, canoeing, boating and sportfishing) has increased considerably, both in areas close to towns and cities and in attractive natural surroundings (the mountains, archipelagos, coasts, river valleys, areas around lakes, etc.). Areas that are particularly attractive for vacation-cottage development are generally those areas of interest from a nature conservancy and outdoor recreation viewpoint.

At present the polluting industries have limited demands because of the prevailing economic situation and ongoing structural changes. The emphasis on a higher level of self-sufficiency in *energy production* will place serious demands on natural resources, such as peat, uranium and wool, as well as the construction of wind-driven and hydroelectric power stations. Concern is expressed about the lower concentrations of *mineral resources* and the more frequent use of open-cast-mining methods. Both these trends can lead to a greater impact on the environment. Alum shale, quartz and ores containing alloy metals are likely to be more and more in demand in the future.

Many demands are placed on the uneven distribution of *freshwater resources,* primarily for agricultural irrigation which is expected to increase twofold by 1985. With regard to *marine resources*, it is primarily fishing that comes into conflict with activities threatening the marine environment. This is particularly the case in the Baltic Sea which is extremely sensitive to increased fishing activities, the expansion of its low-oxygen bottom waters, increases in its high content of heavy metals and other toxic substances, and increases in the shipping of oil and other dangerous goods.

According to this 1979 governmental report, future demands on natural resources require that emphasis, in future national physical planning work, be put on the following eight topics:

(1) Restricting environmental pollution: Great constraint should be exercised in permitting further strains on the natural environment.
(2) The protection of water resources: Conscious efforts should be made to introduce measures intended to conserve freshwater and marine resources.
(3) Increased vigilance concerning general environmental developments: Increasing attention must be paid to the impact on nature conservancy, outdoor recreation and the protection of ancient monuments by environmental changes, caused by rationalization in agriculture and forestry.
(4) Greater emphasis on outdoor recreation and alternative types of vacation housing: Land should be reserved for outdoor and vacation housing in such a way that the attractive areas become accessible to as many people as possible.
(5) Preparatory action concerning the introduction of new energy sources: The increased utilization of domestic energy sources will demand their conservation. Consideration should also be given to other land uses and demands for a good environment.
(6) An improved overview of resources for industry and their conservation: Natural resources of particular value to industry and mining activities should be protected so that future utilization is not restricted. Conditions for the location of extracting activities should be dealt with on a comprehensive basis and with regard to the natural environment's sensitivity to pollution.
(7) The conscious control of settlements growth and the more important investments in transportation and energy distribution systems: The co-ordination of national physical planning, regional policy planning and regional traffic planning should be strengthened, in order to achieve a settlement pattern and a transportation and energy distribution system that meets employment and service demands, while simultaneously ensuring that conservation requirements, concerning land and water resources, are met.

(8) The simplification of working methods: Guidelines for the conservation of land and water resources should be regulated by new planning and building legislation under preparation. The guidelines should be more widely implemented in ongoing planning and survey work particularly at the municipal level.

The report also identifies several areas with special problems concerning the use of land and water, such as soils that are particularly susceptible to ground-water pollution by nitrogenous fertilizers, difficulty in obtaining adequate freshwater supplies, conflicts concerning marine resources utilization, shortage of sand and gravel, or arable land of value to nature conservancy. In the same way, areas with unexploited resources, such as the production of energy and minerals, are identified.

Decision by the Riksdag

The Riksdag has recently approved a Government Bill based on the 1979 Report; the outcome is that national physical planning will continue but on a somewhat less ambitious level than proposed in the Report. This reduction in the scale of national physical planning chiefly affects the natural resources and environmental fields.

FUTURE NATIONAL PHYSICAL PLANNING

National physical planning has now completed its introductory phase and is no longer in an exceptional position. It must now be integrated with normal, day-to-day, planning work, especially at the municipal level. This may lead to less uniformity with respect to nationwide decisions. However, an interaction between central, regional and local levels will continue to be a characteristic feature.

Guidelines will have to be ratified by the Riksdag and the local authorities will, even in the future, have the main responsibility for actual planning work. The State County Boards will implement the guidelines when approving plans and granting permissions. Should a local authority and a State County Board not agree, it will be up to the government to settle the matter.

However, in the future, there may be a need to provide more comprehensive descriptions of the utilization of land and water areas and the environmental situation in the country as a whole. At central government authority level, the National Board of Physical Planning and Building is responsible for co-ordinating information and experience on the conservation of land and water resources. The recently established Delegation for Marine Resources has a corresponding responsibility for Swedish territorial waters. The National Environment Protection Board has overall responsibility for monitoring environmental changes of national importance. These three bodies should continuously review developments in their field, be responsible for the joint studies that may be required and also provide the Government with suggestions about future, proposed measures.

The new role of national physical planning will certainly facilitate cross-sectoral integration, particularly at the municipal level. The treatment of natural resources and the environment will now depend more on the attitude of individual authorities.

EXPERIENCE OF NATIONAL PHYSICAL PLANNING

National physical planning has had several positive consequences. Municipal-wide, physical planning has started at the local level, with the main purpose to treat the national interests but, in practice, also dealing with questions of a regional and local character. A more or less pronounced integration between physical and socio-economic aspects is being developed. This planning has already had a steering effect on new buildings and certain exploitation projects. Furthermore, national physical planning has been important for the protection of continuous areas, especially along the coastline and in mountain regions. Certain main rivers and parts of rivers in the north of Sweden have also been exempted from development for hydroelectric purposes.

The implementation of the guidelines for the management of land and water resources has engaged all municipalities, and a great number of state authorities and associations, at different levels. The consequence of this massive engagement of public and private organizations has been a better awareness of the need for natural resources management and environmental protection. One major result is that greater consideration is being given to these aspects at the local level.

Another positive consequence of national physical planning is that the knowledge of the land and water resources of Sweden has substantially increased. Much good information is now available on nature conservancy interests, agricultural production areas, qualities and quantities of water resources, mineral resources and sensitiveness of the environment to different polluting substances. By means of this knowledge, we can evaluate now better than before the consequences of different use of natural resources.

However, the step from planning to plan implementation is long. In recent years, resistance to concrete action in the natural resources and environmental fields has grown. For a number of reasons, these interests are now viewed less positively: the national economy has got worse; latent conflicts often come to the surface when attempts are made to convert an abstract plan into concrete implementation; resistance from industries, landowners, etc., has grown. Furthermore, the financial resources available for environmental protection, such as safeguarding nature conservancy areas, are quite insufficient.

There is a need to develop Sweden's national physical planning further. The ecological approach has to be given a more concrete form and made more applicable to physical planning at the local level. The problems of implementation have to be re-emphasized. Further decentralization and popular participation is desirable. And a greater attention has to be paid to problems arising in connection with the multiple use of land or water, for instance the environmental impacts of modern agriculture and forestry.

References

Ministry of Housing and Physical Planning (1979). *National Physical Planning.* Stockholm.

Ministry of Housing and Physical Planning (1980). *Housing, Building and Planning in Sweden.* Contributing to the 41st Session of the ECE Committee on Housing, Building and Planning. Stockholm.

Planning Sweden (1973). Allmänna förlaget, Stockholm.

CHAPTER 7

The Political Economy of Pollution Control in the United States

Paul B. Downing

THE pursuit of environmental quality involves both politics and economics, hence the title of this chapter. The discussion[1] will explore the interaction between politics and economics in the efforts of the people of the United States, through their federal government, to adopt, implement and enforce pollution control laws. This is a complex and difficult subject which, firstly, requires a definition of political economy.

The politics in question concern those who must control emissions, those who must pay for this control and those who benefit from the control. The issue of 'who pays' is central to political debates on pollution control. The owners of the environment are those who have the right to use it without compensating others. To illustrate this point, let us suppose that before there were any government efforts to control pollution, emitting firms were allowed to release unwanted by-products into the environment at zero cost. As others in society become concerned with the quality of the environment, those who would benefit from its improvement (recipients) will seek to prevent emissions. There are at least two methods of accomplishing this goal. One is to recognize that the emitters have the right to release wastes. It is then necessary for the recipients to compensate the emitters for the costs incurred in controlling the release of wastes. There are various problems encountered in this solution. Since environmental improvement generally has the characteristics of a public good, individual recipients will have the incentive to act as 'free riders'. This makes voluntary contributions or individual bargaining impractical in a setting involving so many people. Government, with its power to coerce recipients to pay, can be called upon to solve the problem. The end result is a payment from recipients to the government (perhaps in the form of taxes) and also a payment from the government to emitters to reduce emissions.

As Coase has shown, this distribution of property rights can reach an efficient solution. However, it causes a redistribution of wealth away from recipients and toward emitters. If the level of control chosen is efficient, the recipients would still be better off than if there were no control: their benefits would exceed their payments. However, they would be even better off if they

[1]This paper draws in part on work done jointly with Gordon Brady to whom I am indebted.

did not have to pay. Suppose instead that the recipients were to assert that they had the right to a clean environment, free from the damages done by the emitters. Essentially, the recipients would be arguing that they owned the property rights to the environment. If recipients were successful in this argument, emitters would have to pay recipients in order to be allowed to emit. This distribution of property rights will also lead to an efficient allocation of resources, but recipients will have more wealth than if emitters owned the property rights.

The political debate on pollution control is a debate about who owns the property rights to the environment and, therefore, who pays for pollution control. In essence, recipient lobbies have found it less expensive to acquire property rights to the environment through the political process than through the market. The outcome of this political activity in the United States has been a gradual shift of property rights from emitters to recipients. Currently, the property rights are shared between the two groups. Emitters are allowed to emit some pollutants without paying but must pay to control further emissions. Political debate continues, however, because it will always pay either group to acquire more of the property rights. They increase their wealth by doing so. Furthermore, it pays for each group to defend itself against redistributions of rights away from it because of the wealth it would lose through redistribution.

The shared property rights structure at present employed in the United States can produce an efficient allocation of resources, in the same way as either of the pure rights cases could determine efficient levels of control, and can reach them at least cost. However, there is a feeling among environmental economists in the United States that resources are, in fact, wasted. The regulatory approach, it is argued, produced inefficient outcomes. In particular, environmental quality is produced through inefficient means and, therefore, produced at an excessive cost. If this is true, it is logical to ask why. The answer is that political incentives cause us to choose these inefficient means. The important incentives in this case are those faced by individuals. Individual sources of emissions will attempt to reduce the level of control of their emissions. In this way, they can avoid costs. To an extent, they are acting as free riders. Suppose that the owner of an emitting firm is concerned about environmental quality. Since the firm's emissions are only a small percentage of total emissions in its location, reductions in control are perceived to have very little effect on environmental quality. But the cost savings are purely private. As discussed later, the individual case-by-case approach of the US regulatory system allows emitters to reduce control and spread the environmental costs over a large group of people.

INCENTIVES OF MAIN POLITICAL ACTORS

Political Entrepreneurs

A politician acting in his own self-interest will seek to adopt laws which will increase the probability of being re-elected (or elected to higher office) (Becker, 1976; Downs, 1957; Freeman and Haveman, 1972; Peltzman, 1976). A politician's goal function may also include his desire to forward his view of the 'public interest'.[2] The politician's ability to pursue his goal function depends upon his power, not just his re-election. Power is the function of his tenure in office and his level of office. Policies which confer benefits on constituents will be supported, while policies which are costly to supporters will be avoided if possible. Politicians will seek to develop policies where the benefits are easily understood and the costs hidden or borne by others (Tullock, 1967).

A politician's 'constituency' will differ depending upon the level of government being studied and upon the aspirations of the particular politician for higher office. The ability to hide costs or shift them on to others will also differ with the level of government. Federal-level politicians will attempt to pass on the costs of their decisions to state agencies and politicians. State politicians, on the other hand, will find it in their self-interest to resist such attempts.

Bureaucrats

A government bureaucrat acting in his self-interest will seek to increase his income and power. This will be true whether or not his goal function contains as a major argument the goals of the agency for which he works (Niskanen, 1971 and 1975). Both income and power can be generated by increasing the responsibility and discretion of the bureau. Increased responsibility provides the basis for a larger budget and budget is positively related to income and power. Quite obviously, it will be in the bureaucrat's self-interest to be an active participant in political decisions affecting his agency. An agency's legislative mandate directly affects the responsibility and discretion of the bureau and, hence, the income and power of the bureaucrat. Consequently, he will advocate policies which will favour his bureau with either more responsibility or discretion.

Firm Managers

In many ways, the goal function for the manager of a firm is similar to that

[2] In the Frey and Lau (1968) model, the ideological variables become feasible behavioural strategy only when popularity is high, say for an expected vote of 52 per cent. When popularity is low, attempts are made to manipulate policy variables to increase the number of votes obtainable.

of the bureaucrat. He/she wishes to increase his income in real terms. He/she derives increased income from making the firm more profitable. The manager may have other arguments in his goal function, such as social prominence and social responsibility. However, his ability to pursue these other goals depends significantly upon his success in providing a rate of return which is competitive with similar investments (Williamson, 1964). Profits may be increased by investing resources in obtaining favourable interpretations and/or changes in pollution control laws. One might expect the same type of rational cost-benefit calculus for the political investments of firm managers as would be undertaken for capital expenditure.

CIG Bureaucrats

The Sierra Club is one example of an organization of citizens concerned about environmental quality. Such organizations are called Citizen Interest Groups (CIG). The people who become involved in CIGs might appear to be beyond the scope of our self-interest model but we do not believe this to be the case. We believe that CIG leaders are attempting to pursue an individual goal function of changing resource allocations to better reflect their view of the social ideal. They can do this by changing the *de facto* property rights of the system. Like bureaucracts and firm managers, CIG leaders are faced with alternative means of obtaining their goals. They will weigh the expected costs and benefits of alternative actions in order to choose that combination which will maximize their utility.

It is in the CIGs' self-interest to seek changes which decrease their costs of obtaining decisions. Efforts to increase the costs of decisions favouring opposing views would also be consistent with CIG self-interest. The cost of obtaining a judicial precedent may be high but it is probably less than the cost of obtaining the same decision through legislative action. Convincing one judge or a three-judge panel (at the appellate level) would appear to have a lower cost than securing the votes of a majority of the legislature. The costs fall if the judge shares similar views about environmental media. Consequently, quasi-legislative activity through the courts has been a cost-effective means of establishing rules favourable to CIGs (Goetz and Brady, 1976).

Summary

In summary, the institutional incentives and goals of the individuals holding various public and private positions lead to a complex set of mutually beneficial trades. At each point in the political bargaining process, benefits are conferred on some and costs imposed on others. Furthermore, based on the self-interest paradigm, each actor may be expected to seek out those exchanges which offer him/her maximum net benefits. The timing and incidence of costs and benefits may be adjusted through trades. Since the political bargaining process may be used to alter the costs of decision-making by other actors, it is rational for the various decision-makers to want to increase the costs of making decisions which are likely to be unfavourable to their goals and to reduce the costs of making decisions which are generally to their benefit. The influence of any one group will differ among issues and for locations. Groups may adopt different policy perspectives at different locations due to changes in the weights of arguments in their goal function or in their constraint set. Coalitions will form around specific issues and dissolve when those issues have been settled. Members of a coalition may find themselves in opposing coalitions over the next issue or at another location. The complexity of the bargaining process can be reduced to understandable dimensions by applying our constrained self-interest paradigm.

Both the passage of a law and its implementation offer opportunities for gain by various groups. In order to further develop our understanding of the effects of constrained self-interest on public policy formation, let us explore the passage and implementation of the federal Clean Air Act of 1970.

PASSAGE OF THE 1970 AMENDMENTS

The debate and passage of the 1970 Amendments to the Clean Air Act took place in an atmosphere of urgency to solve the air pollution problem. From its passage, the 1967 Clean Air Act became the subject of critical scrutiny by Congress. Oversight hearings to assess progress under the Act were held in 1968 and 1969 by Senators Edmund Muskie and Thomas Eagleton and Congressman Paul Rodgers. The hearings showed that progress was indeed slow (Bonine, 1975).

There was growing dissatisfaction with 1967 air pollution legislation which relied upon voluntary state action. The environmental movement was gaining momentum and its principal actors had learned that the existing institutional structure failed to encourage state agencies to be responsive to their demands. The air pollution issue provided a convenient vehicle for politicians to build constituency support. At that time, California (a key electoral state) was the grass roots of the environmental movement. Los Angeles had the worst oxidant problem in the nation. The Los Angeles Air Pollution Control District (LAAPCD) had come under vigorous attack by environmental CIGs for perceived deficiencies in the control of emissions, particularly those from oil refineries. The CIGs viewed the LAAPCD as a captive of the firm it was under mandate to regulate. CIG efforts to force the California Air Resources Board (CARB) to take a more activist stance in Los Angeles were unsuccessful. With limited success at the state level and disillusioned by their prospects, CIGs turned to the federal politicians for help. The result was the passage of the 1970 Amendments to the Clean Air Act.

The 1970 Act set up a procedure designed to guarantee clean air by 1977.

The Environment Protection Agency (EPA) was to set national air quality standards. The states were required to develop and implement plans, called State Implementation Plans (SIP), to meet those standards and to do so by a 'date certain'. If they failed, the EPA would take over. This provision became known as 'action forcing'. The 1970 Act also provided for citizen enforcement so that the EPA could not avoid the takeover requirement.[3]

The mechanism adopted in the 1970 Act effectively allowed costs to be shifted from the federal politicians to the states and EPA. In proposing the state implementation plans, the officials and politicians at the state level would incur substantial political costs. However, virtually no costs would be imposed on Congress or politicians at the national level. This feature was conducive to CIG supporters since they were seeking to impose these costs on the states. The action-forcing mechanism, with its date certain features, created a federal omnipresence which made it more expensive for a state agency to avoid its duty under the Act. The discretion of state and local agencies was circumscribed by the setting of federal standards and the requirement that states submit acceptable SIPs to the EPA.

Bureaucrats at the federal level stood to be the net gainers from the new law. The 1970 Act represented a major shift in emphasis as compared to its antecedents. It was no longer necessary for the EPA to wait for the states to produce air quality standards and then come up with suitable control strategies to meet the standards. Failure of a state to act expeditiously placed the federal government in the position of being able to pre-empt state control activities. The law gave the federal agency many tasks which had previously been vested in the states. Obtaining additional responsibilities permitted an expansion of the budget, employment levels and an augmentation of power for federal bureaucracy.

The National Air Pollution Control Administration (NAPCA) was the precursor agency to the EPA. John Middleton, in his role as head of NAPCA, actively lobbied for an increased federal role. Even before debate on the 1970 Amendments was initiated, Middleton was attempting to build a constituency supporting stringent air quality standards in the development of State Implementation Plans (SIPs) and, ultimately, to support the federalization of the control effort. To these ends, grants were made by NAPCA to various national organizations to conduct air quality workshops in major cities during 1969 and 1970 (Sabatier, 1975).

The net effect of the NAPCA programme is not clear. But its true intent comes through in the testimony at the 1972 Oversight Hearing by Frank Partee, Technical Director of the Kentucky Air Pollution Control Commision. He argued as follows (US Senate, 1972): "As the time for public hearings approached, this NAPCA involvement with citizen groups, which earlier had been so constructive, took on another colour. Representatives were sent by NAPCA, out of its Washington Offices. These representatives often appeared without knowledge of State officials and even without knowledge of NAPCA's

[3]Clean Air Act 1970: Section 304 (a) (1), 42 USC, Section 1857h-2 (a) (1).

own regional offices. They carried a strange message. They said, 'Don't trust your State agency. It won't act in your best interest.' People were coached on how to present testimony and opinion was prejudiced against State proposals even before they were publicly released."

NAPCA even took its argument to Congress. Middleton had some of the NAPCA-funded CIGs send telegrams to Senator Muskie during the hearings prior to the passage of the 1970 Act. The telegrams complained of CIGs having been excluded from state hearings on the setting of air quality standards.

As indicated, a federal control effort, placing substantial pressure on state agencies to control more efficiently, would certainly be politically palatable to federal politicians and CIGs. CIGs received more effective means of exerting pressure on the state and local control agencies. Federal bureaucrats were, of course, not indisposed to the significant increase in the budget and power which accompanied the expansion of their role. Similar motivation might explain state bureaucratic acceptance: a strong federal push meant increased state budgets. Firms could expect to lose from the passage of such an act. However, since the real decision-making powers affecting firms remained at the state and local level, they apparently thought that mitigation of the adverse effects would be possible through bargaining at these levels.

IMPLEMENTATION OF THE 1970 ACT

One of the first events required by the action-forcing mechanism of the 1970 Act was the proposal of federal air quality standards. The EPA proposed an oxidant standard of 0.06 ppm, the level suggested by NAPCA prior to the passage of the Act. This proposed standard met with considerable opposition, particularly from the California Air Resources Board (CARB), who argued that there were no demonstrated health effects below 0.10 ppm.[4]

Why, then, was the 0.06 figure chosen? Our model yields substantial insights into the emergence of this decision. First, recall that prior to the passage of the 1970 Act, Middleton was attempting to build a constituency for a federal role through NAPCA-funding of CIGs. In order to gain the support of California-based CIGs, and to put pressure on the CARB to adopt more stringent air quality standards under the 1967 Act, NAPCA had to look tougher than the CARB. It makes sense that, in order to accomplish this, Middleton advocated a tougher air quality standard than the one already in effect in California.

By previously committing the federal agency to a stringent standard, Middleton made it difficult for the EPA to back down without losing the constituency he had built. Furthermore, the 1970 Act created an additional

[4]Interviews with John Maga, former Executive Director of the California Resources Board, on 1 April 1975, and with Robert G. Lunche, Head of the Los Angeles County Air Pollution Control District, on 3 April 1975.

incentive to be tougher than California. If the EPA adopted a standard less stringent than California's, EPA's power to effect control policy in California would be significantly reduced. The potential loss of control in California also explains the consistently pessimistic assessment of technical options by EPA analysts. Given the action-forcing mechanism, the more pessimistic the EPA's stance on the effectiveness of control effects, the more control would be required to meet any standard. Hence, the EPA could obtain the power to dictate the state's control policy. The ability to reject SIPs and the pre-emptive federal threat represented a significant augmentation of the EPA's bargaining power. By defining the problem pessimistically, the EPA increased the scope of their bargaining power with a state. Their continued contractual and grant support of the CIG-participation programme throughout the implementation process was a key element of their bargaining strategy.[5] The promotion of the Clean Air Act as a binding legal constraint by CIGs under contract to the EPA through their educational programmes, the media, and litigation increased the political viability of EPA policies at the state level.

Through the passage of the 1970 Act, the EPA acquired real power to affect resource allocation. The power to impose costs on politically powerful groups led to efforts to constrain its use. President Nixon sought to check and limit the EPA's use of power through the Office of Management and Budget (OMB) review of EPA regulations. Ruckelshaus attempted to retain this power but was ultimately unsuccessful (Quarles, 1976). Senator Muskie and others attempted to obtain the same type of controls over the EPA. They met with some success, primarily because of their power over legislation and funding. It is important to note that Senator Muskie, in addition to his position on the Public Works Committee, was chairman of the Budget Committee. A third group, the CIGs, were also engaged in attempts to curtail the use of discretion by the EPA and the states. In the case of the *National Resources Defense Council (NRDC) versus the EPA,* the court ruled that sources and the states cannot agree to individual changes and delays (variances) in their emission control after the EPA has approved an SIP (Case Ref. No. 5 ERC 1879). In another case, the NRDC was successful in reducing its costs by forcing all court reviews of SIPs to be held in the District Court (DC) Circuit. This accomplished another goal since this court had traditionally been more sympathetic to the environmental cause than had other circuit courts.

The EPA's efforts to establish grounds for, and constraints on, permissible bargaining with the states were abruptly halted in their early stages by CIG action. Shortly after the 1970 Act's passage, CARB adopted the attitude that the law was totally unreasonable and CARB would not ask the people of California to comply with it. CARB submitted an SIP which argued that the air in Los Angeles would never be as clean as the federal air quality standard.

[5]The funding of Citizen Interest Groups (CIG) through contracts and grants began in 1969 with the National Air Pollution Control Administration. It was pursued by the EPA Office of Public Affairs. For the period 1972-1975, some CIGs received as much as US$35,000 per year to promote EPA policy alternatives through the media, the courts and the legislature.

The EPA had no choice but to disapprove the SIP. However, they gave CARB until February 15, 1973, to submit a revised SIP. Disapproval of the initial SIP submitted by California shifted the responsibility for correction of deficiencies to the federal level. In *City of Riverside vs Ruckelshaus,* a case brought against the EPA by Los Angeles area CIGs, the courts held that the EPA had failed to perform a non-discretionary duty and must adhere to the statutory requirement to promulgate an SIP for Los Angeles (DC Circuit California 1972. Case Ref. No. 4 ERC 1728). The court ruling meant that EPA threats were no longer a viable means of 'encouraging' the states to produce a more effective plan. The court required the EPA to produce an acceptable SIP. In meeting the court order, the EPA produced a plan which required stringent control of mobile and stationary sources and over 80 per cent gas rationing. After the EPA proposed the gas-rationing plan in 1973, CARB's opposition became politically viable because the incidence of the costs of EPA policies became clear to the residents of California. The political pressure was removed from CARB and placed instead on the EPA.

CIGs were seeking to increase their property rights to clean air by the imposition of control costs on a reluctant CARB and LAAPCD. The result of *Riverside* case was the opposite. Instead, those costs were imposed on the EPA. In *EPA vs Brown,* the court ruled that the CARB need not enforce the federal SIP (Case Ref. No. 7 ELR 20366). Because the costs of their plan were politically unbearable, the EPA chose to openly violate the requirements of the Act. CIGs realized that they were powerless to engage either the state or the EPA in further litigation to require implementation and enforcement of the SIP, even though the EPA was in violation of the mandate in *Riverside.* The 1970 Clean Air Act was dead, killed by the self-interest which had shaped its creation. The decision-making power was returned to the state, where it had been prior to the passage of the 1970 Act.

Passage and implementation of the Clean Air Act Amendments of 1977 did not solve the problems of the 1970 Act. Attempts were made to again shift the political costs to the states. Again, the states were successful in avoiding most of these costs. As a result, Congress is currently debating the Clean Air Act Amendments of 1981. It can be expected that the Reagan Administration and Congress will formally return to the states much of the power to decide on environmental policy.

IMPLEMENTATION ISSUES

The implementation problems outlined so far are not unusual. In fact, they are the issues commonly found in all implementation studies in the United States, regardless of the law being implemented. In the following discussion, we shall identify four basic findings of implementation studies. The details surrounding these findings will differ greatly among issues, but the general behaviour found is consistent over a wide range of policy issues.

Technical Issues

Lawmakers have adopted the philosophy that policy issues can be stated in purely technical (scientific) terms and can be solved by purely technical analysis. Thus, as we have seen, the air pollution problem, as defined in the 1970 Act, is purely a technical one. The goal is defined as zero health effects, an obviously technical notion. Control is also defined in technical terms: add controls to individual sources until clean air is attained. The implicit assumption is that technical assessments are valueless. Scientists measure health effects and the findings clearly define where 'zero' is. Technical issues are outside political and economic considerations—not so. In the US, the Army Corps of Engineers has been required to conduct a benefit-cost analysis of each proposed project and to pursue only those with positive net benefit. It has become common knowledge that the Corps analysts systematically overestimate the benefits of a project and underestimate its costs, in order to increase the number of projects the Corps can justify and build.

The Corps is not the only agency which manipulates technical analysis to produce politically acceptable, or hoped for, results. There are many cases in air and water pollution as well. It must be realized that technical knowledge is not absolute and certain. There is almost always a range of uncertainty on any technical issue. When the possibility of different assessments exists, it is then possible for agencies to choose an interpretation of the data which favour their preferred position.

Constraints On Policy Implementation

The output of an agency is the result of many factors. Laws, such as the Clean Air Act of 1970, outline policies to be implemented by an agency. In order to accomplish this task, the agency must use the resources available to it. These resources include personnel, budget, legal authority and time. Typically, the resources provided are insufficient to accomplish the assigned task. Hence, the lack of availability of sufficient resources of several types constrains the achievement of policy goals.

We find that the existence and form of a pollution control law are not the only constraints operative in its implementation, nor need they be the most significant. Other influences, such as resource constraints and economic interest in the *status quo,* constrain the agency in its pursuit of the goals of the law. One additional point, raised earlier, should be emphasized. Not only is it true that other constraints on agency policy exist, but also the law does not provide a completely effective constraint on agency behaviour. Provisions of a law can be ignored and even court orders can be evaded in many cases. Thus, we found that while the EPA was 'forced' by the courts in the *Riverside* case to produce an effective SIP, it complied in such a way that the order was not really effective. This observation leads on to the next point.

Bureaucracies are not necessarily interested in attaining the legally mandated goal of the agency

The naive idea that agency bureaucrats have, as their sole goal, the furthering of the intent of those who passed the law that the agency is charged with implementing has been discredited by policy analysts from both political science and economics. Downs (1967) provides a listing of potential motives or goals of bureaucrats: power, money income, prestige, convenience, security, personal loyalty, pride in proficient performance of work, desire to serve the public interest as defined by the bureaucrats and commitment to a specific programme of action. He assumes that every official acts, at least partly (and sometimes only), in his/her own self-interest. The first five of the goals just listed are characterized by Downs as self-interest goals, while the other goals have both self-interested and altruistic elements.

A result of the action of self-interest bureaucrats is that they manipulate information so as to reduce public intervention in their activities and to reduce public accountability for failing to achieve legally specified goals. (A 'presumably' secondary effect of the LAAPCD's classification of information as confidential was to reduce the ability of other agencies and the public to determine exactly what they were doing.) Inability to determine agency output insulates the agency from criticism. Likewise, creating barriers to understanding makes it difficult for the public to determine the true effect of policy. Thus, we can see the advantage to the agency of defining policy issues in highly complex technical terms. In this way, it is difficult for those without specific technical training to understand what is being said or done. And finally, agency bureaucrats are not the only ones who manipulate information and make it technically complex.

Policy Outcomes

The self-interest paradigm suggested as a model of bureaucratic behaviour is easily adapted to an analysis of the other parties involved in the implementation process. Several writers have characterized the implementation process as the result of complex bargaining among self-interested individuals and groups. Bardach (1977), for example, characterizes the implementation process as "a process of strategic interaction among numerous (groups of) special interests all pursuing their own goals, which might or might not be compatible with the goals of the policy mandate." These groups are politicians, bureaucrats, firm managers (and other emission-source managers) and citizen interest groups. Each type of group is represented in policy debates at the federal, state and local levels but not necessarily by the same organizations or individuals. Thus, the special interest represented at the local level may include some specific individuals or groups also active at the federal level and some individuals or groups that are only active in the particular local issue being debated. Each individual or group joins the debate on a specific issue when he/she perceives that it is in his/her self-interest to do so. This will be the case when the perceived benefits of action on this issue exceed perceived costs. Since benefits of action differ among issues, and among individuals or groups, and since the costs of action also differ, we find the composition of the bargaining groups will differ among issues. Consequently, so will the outcomes. Implementation policy is the result of a series of policy decisions at the federal, state, and local levels, each characterized by bargaining among those individuals and groups who find it in their self-interest to act at that level on that policy issue.

At the federal level, it is noted that the policy issues are most often general in nature. The setting of air quality standards is a good example. The standard does not say who shall reduce their emissions or by how much. Thus, the cost of the policy to sources is unspecific, and therefore, the sources perceive no direct benefit to bargaining at this level of policy. The CIGs, however, perceive direct benefits from increasing the stringency of the standard. The result is an abstract and idealized national air quality standard.

As policy issues proceed from general to specific, the benefits and costs of policy bargaining change for each group. When the issue is national industry-wide emission standards, sources receive higher benefits from bargaining. However, at this level the issue is still fairly general. Therefore, the benefits of bargaining for an individual source are not great. Instead sources co-operate, typically through an industry association, and bargain jointly. At the same time, the benefits of bargaining on any single industry emission standard are reduced for the CIG. This is because this emission standard is only one of hundreds which affect the environment. Also the cost of bargaining by the CIG increases because the issues are now much more technical. In order to be effective, the CIG must be able to show that other more effective technologies are available and could be used effectively in this industry. The inevitable result is a weakening of individual industry emission standards. As the policy issues become even more specific, the benefits of bargaining increase for affected firms while the costs of bargaining increase for CIGs. This observation implies that local industry will be able to further reduce the stringency of the regulations with which it must comply. This has been shown in various enforcement studies (see contributions by Brady and Bower, Downing and Kimball, and Ullman, all in Downing and Hanf (Eds.), 1981).

MODELLING IMPLEMENTATION

In order to understand the implementation process and how changes in the institutions and laws affecting it will alter the outcome of policy deliberations, we have developed a model of implementation (Downing,

1981). We assume that all individuals active in decision-making on pollution control are guided in their choices by their own self-interest. In other words, they seek choices which make them better off directly and, in addition, they seek choices which improve the social interest as they perceive it.

The individual is constrained in making choices by income, wealth and time considerations. In attempting to maximize self-interest, the individual considers both the cost (monetary and time) of each choice and the likely benefits (improvements and self-interest) of making that choice. The individual will choose those options which yield higher benefits per unit of costs. (Technically, the individual will choose combinations which will maximize perceived net benefits, given his/her wealth and time, and the relative costs of other choices.)

Control Agency Manager

There are many possible arguments in the goal function of the control agency manager, two of which will be discussed here. However, it should be understood that we do not imply that these are the only arguments, only that these two are useful for the analysis to follow.

Niskanen's original argument (1971 and 1975) was that the bureaucrat seeks to maximize his or her budget. This goal was subsequently changed to the maximization of discretionary budget. We will adopt this as one of the bureaucrat's goals. Note that there are many ways a bureaucrat may use discretionary budget. Discretionary resources can be allowed to further other goals of the bureaucrat. In a multiperiod model, discretionary resources can be allocated for this period to affect the bureaucrat's input parameters in future periods. This feedback can be very important. The second goal we attribute to the bureaucrat is improvements in environmental quality. It is certainly true that at least some of the people who have been assigned to top positions in federal pollution control agencies have previous records and substantial national reputations as being 'pro-environment'. Many others within the control agency also will favour improvements in environmental quality.

The implications of these two goals can now be explored. Under the usual utility maximizing assumptions, the bureaucrat will be willing to trade-off some of one goal for an amount of the other goal and remain at the same utility level. Such a utility function is shown in figure 1, Panel A, where DA is defined as a vector of discretionary activities and EQ is defined as a vector of environmental quality. EQ_t increases to the east (right) and DA_t increases to the north (upward). For the purpose of this discussion, the initial presentation will assume that an equilibrium exists in each period and that the equilibrium values of DA and EQ are the same for each period.

Note that it is not strictly necessary for the agency manager to prefer both vectors of goods in a multiperiod model. Suppose that the agency bureaucrat is indifferent to both present and future environmental quality. The bureaucrat is a pure maximizer of the present value of discretionary activity. In the

Figure 1: Static Equilibrium Model of Pollution Control Implementation

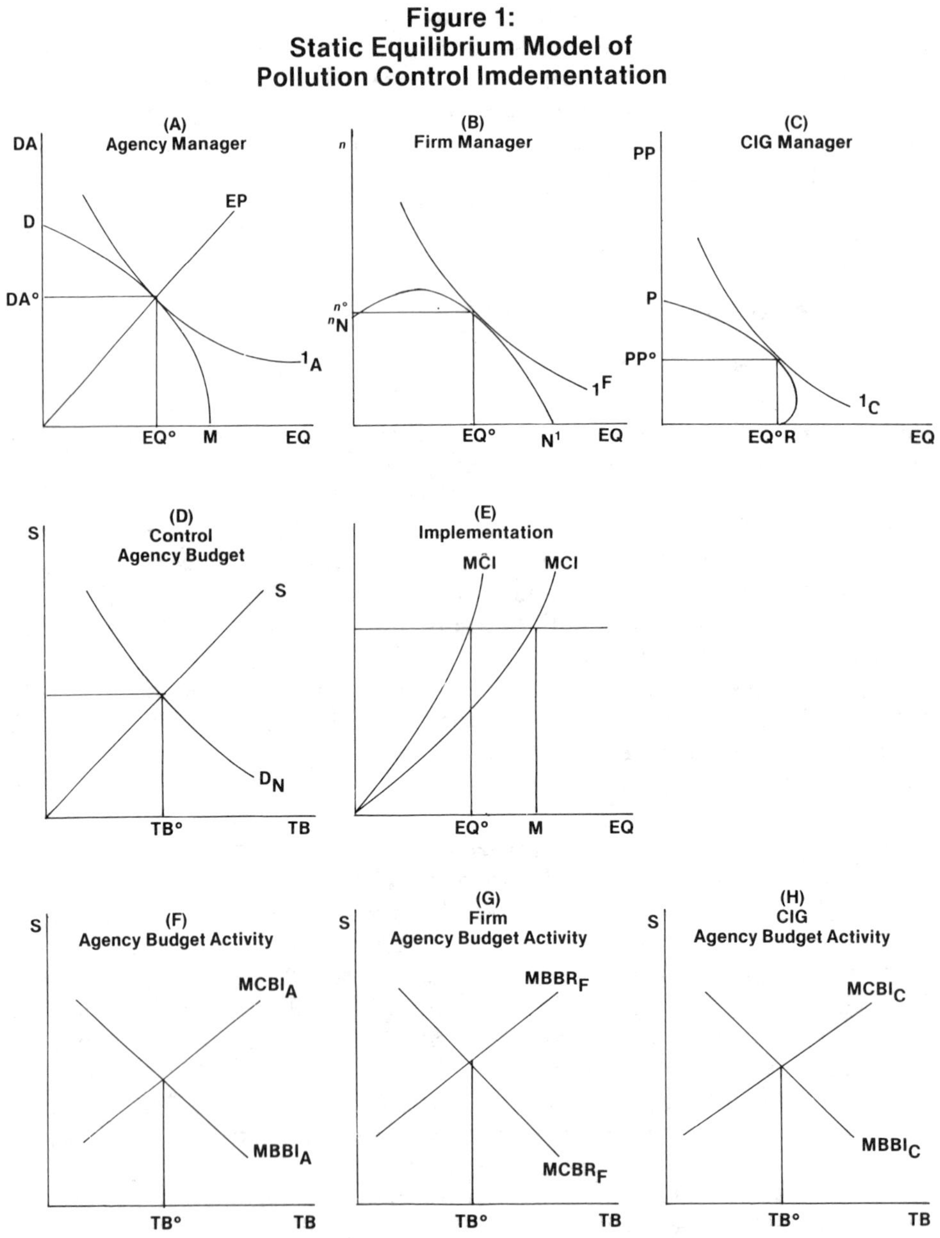

absence of feedbacks in future periods, maximizing DA in each period will maximize its present value. This being the case, the bureaucrat's utility surface in any period would appear as horizontal lines increasing to the north. However, if input parameter values (and therefore the DA attainable) in future periods depend in a direct relationship upon the EQ he or she provides in this period, the bureaucrat's purely discretionary activity maximizing behaviour over the multiple periods would include positive (derived) preferences for EQ in early periods. It is also possible that the bureaucrat will be indifferent during any period to discretionary activity. (He is a pure present value of environmental quality maximizer.) Again, the affect of discretionary activity during this period on inputs in subsequent periods must be considered. If there is a positive relationship between this period's discretionary activity and next period's input parameters, the bureaucrat will have a (derived) preference for DA in early periods.

In seeking to maximize beyond his or her preference function, the bureaucrat faces a budget constraint. This budget constraint is derived from available inputs, cost functions, and prices. The agency manager does not produce his or her consumed environmental quality directly. It is the emitters who control emissions and thus control EQ. The agency manager can influence their choices by using the available inputs to produce intermediate outputs, which in turn affect emitter choices. At any point in time the inputs available to the agency manager include the existing set of laws, general administrative regulations and court interpretations of those laws; the agency budget, including any constraints on the use of funds, such as required grants or research; personnel; and information on the state of the world. Information includes knowledge or estimates of the actual condition of the environment, source emission controls and the expected reactions of the other groups to agency policy. The intermediate outputs which these inputs can produce include: regulations on individual source emissions and general environmental quality; agreements with emitters on the level and timing of control; citations and court action for violations of the regulations; information on the state of the world and the agency's position on important policy; and spending on discretionary activities of various sorts. These intermediate outputs combine with activities and preferences of the other groups to produce the final outputs received by the agency manager.

The agency manager faces a production function and associated cost function for the final output EQ. If the agency manager used all the available resources efficiently to produce EQ and ignored the effects on next period's input availability, there would be a least cost combination of intermediate outputs which would produce any desired EQ. The resultant cost curve for the agency can be termed the marginal cost of implementation (MCI in figure 1, Panel E). This cost curve depends upon the agency's production function. It, in turn, depends upon the reactions of the other two groups to agency policy. For example, if the emitter decides, for whatever reason, to increase his or her resistance to agency regulations through court action, MCI will shift upward to the left.

MCI assumes that all agency resources are efficiently employed to produce EQ. As we have pointed out, the agency manager may wish to use less resources pursuing this goal in order to obtain some DA. The total budget (TB) available to the agency can be expended on discretionary activity or environmental quality. Suppose that the agency had a total of TB* for a period. If the agency spent all of its discretionary budget (DB*) on producing EQ efficiently, it could produce a quantity M of it. For this given DB*, when less EQ is produced during the period, more budget is available to finance discretionary activities. These funds, when divided by the price per unit of DA, determine the number of units of DA which the agency manager can produce given DB*. Thus, a budget constraint can be derived (DM in figure 1, Panel A). Note that DM depends on MCI which in turn depends on the reaction of the other groups.

As TB changes, DM will change. The agency manager's utility maximizing allocations can be described by the income consumption expansion path EP which is drawn in figure 1, Panel A, as a straight line, though it need not be. An agency environmental quality supply function now can be derived ($\hat{MC}$I in figure 1, Panel E). This function will lie upward to the left of MCI because of the inclusion in this function of discretionary spending. The other groups do not observe $\hat{MC}$I. Instead they observe a point on MCI. These groups may have some idea that the agency manager is spending some resources on DA but it is costly to determine the exact amount and it is also costly to affect this level in present and future periods.

We can now turn our attention to the last component of the agency manager's section of the model; the demand for agency budget. The agency manager can reach higher levels of utility as the budget increases. Under usual diminishing marginal utility assumptions, the bureaucrat's benefits for budget increases would increase at a decreasing rate. Figure 1, Panel F, depicts the agency manager's marginal benefits of an agency budget increase ($MBBI_A$). The horizontal axis represents the total budget of the agency (TB).

In order to obtain additional budget, the agency manager must expend some discretionary resources to prepare an argument for more funds. This argument would be presented before executive and legislative budgetary authorities. The marginal cost of obtaining a budget increase ($MCBI_A$) will be increasing as shown in figure 1, Panel F, for several reasons. The agency manager has other methods of increasing the quantity of EQ or DA. These can include more court action, more support of others who desire more EQ and cost reductions to those who control emissions. As expenditures on direct efforts to affect the budget increase, the opportunity costs of additional funds in terms of improvements in the agency manager's utility increase. Another reason for this cost increase is that the resistance of other interests, who also seek funding for their agency, will increase as more funds are allocated to the control agency. In other words, the executive and legislature face increasing opportunity costs of providing more budget to the agency. Finally, the position of $MCBI_A$ depends upon the level of effective resistance to budget increases generated by the emitter group and the level of effective support for budget increases generated by the pro-control group. $MCBI_A$ is drawn holding these two efforts constant for the period. It represents the agency manager's marginal cost of being effective at increasing his or her budget, *ceteris paribus.* The intersection of $MBBI_A$ and $MCBI_A$ represents the budget equilibrium for the agency manager. It implies that the agency manager will find some finite level of budget optimal.

Firm Manager

Anti-control pressures are generated by the sources of emisions. The reason for this pressure is that emission control is costly and this cost reduces the emitter's welfare. Taking an individual firm and its manager as representative of the group, we can develop the anti-control portion of the model.

We assume that there are two relevant arguments in the firm manager's utility function: the firm's profit and environmental quality. The profit argument is included for all the usual reasons but especially for its effect on the firm manager's income. The existence of EQ in the firm manager's preferences has at least two justifications. First, firm managers have characteristics similar to those found for the typical Sierra Club member. Thus, either directly or through peer-pressure, the firm manager would prefer more EQ. Secondly, there exists a potential feedback in the model which can generate a derived preference for EQ similar to the bureaucrat's derived demand. Co-operation on control efforts this year may tend to reduce pressure for more control in the future. The present value of firm profit may be higher if the firm's manager acts to control, at least to some extent, during early periods. For at least these two reasons, the firm manager will have a willingness to forego some profit this year, and by implication some monetary income, in exchange for an increase in environmental quality generated at least in part by control of the firm's emission. Such a trade-off is shown as I_F in figure 1, Panel B.

The trade-offs available to the firm manager between profit (π) and EQ are the result of several factors. The EQ intercept (N in figure 1, Panel B) is the maximum control possible at zero profit for the period. Attempts to control beyond N would cause the firm to close in the long run. This maximum is dependent upon the firm's cost of emission control and the competitive position of the firm. As the firm controls less, it saves pollution control costs. As control is lowered (moving to the left on the EQ axis), given a level of agency budget, laws, court decisions, etc., the probability of being fined for excessive emissions will increase (see Downing and Kimball in Downing and Hanf, 1981). At some point, the savings in control costs from a marginal reduction in control will be offset by an increase in the present value of probable fines. This is the point where π would be maximized, *ceteris paribus.* Further reductions in control will lead to lower profit. The zero control intercept π_N represents the profit available at zero control. It need not be positive. The firm manager will maximize his/her utility, given agency and other group efforts, where I_F is tangent to the profit/environmental quality trade-off surface $\pi_N N$. The firm manager does not produce the two final outputs included in his/her goal function. Like the agency manager, the firm manager uses available inputs to produce intermediate outputs. These intermediate outputs, interacting with others in society, generate a movement towards desired final outputs.

The influence on the agency through the budget is typical of the firm manager's options. The firm manager has a demand for an agency budget reduction. With lower budget, the agency will be less able to regulate and enforce emission controls effectively. This will shift the firm manager's trade-off surface upward, for example to π_N N in figure 1, Panel B. The firm manager will be on a higher indifference curve. Assuming declining marginal utility of income, the marginal benefits to the firm manager of reductions in the agency budget are reduced as a lower agency budget is achieved. Such a curve is shown in figure 1, Panel G. Remember that the total utility increasing direction of agency budget change for the firm manager is to the left rather than to the right.

The firm manager can affect the agency budget through various direct and indirect means, including the production of political pressure. The marginal cost of an effective campaign to reduce the agency budget ($MCBR_F$) will be high when the agency budget is low and will become lower as the agency obtains more budget. This is true for at least two reasons. First, the firm manager has allies in the other groups seeking government funds. Thus, at high agency budgets, a small additional reduction pressure from the firm will be added to an already high resistance from others. Second, the ability to argue that the costs of meeting requirements are excessive, and the desirability of controlling further is doubtful, will be greater at the higher levels of control implied by a large agency budget. Consequently $MCBR_F$ will slope downward to the right as in figure 1, Panel G. The interaction between $MBBR_F$ and $MCBR_F$ represents the equilibrium agency budget for the firm manager. It implies that the efficient agency budget as viewed by the firm manager is positive rather than zero given the existence of the law, *ceteris paribus.* Finally, note that this is true even if the firm manager is a pure short-run profit maximizer. He/she would then have horizontal indifference curves. Lower budget would allow higher profit but, so long as the law is enforced, the profit maximizing pollution control is positive, not zero.

Citizen-Interest-Group Manager

The third group to be included in the model is a representative of pro-environmental control pressure groups. We will consider a single citizen-interest-group manager as representative. There are at least two arguments in the CIG manager's utility function: environmental quality (EQ) and political power (PP). Both of these preferences may be directly held by the CIG manager but either could be derived from the feedbacks in the system. The desire of the CIG manager for EQ may be derived in part from the membership of the CIG. These members support the CIG financially. They expect the CIG manager to obtain improvements in EQ. If the CIG manager is unsuccessful, they may withdraw their support or remove him/her from office. The CIG manager's desire for political power may derive from hopes of future political or governmental office or from the pure fun of being able to exercise it. However, PP is also potentially valuable in effecting policy choices of the agency and the firm.

Giving up some EQ during this period to obtain more PP may enable the CIG manager to obtain more EQ in future periods. This is because PP obtained during this period will be retained to some extent next period. However, PP can be expected to deteriorate with time. Even if the CIG manager only wished to maintain a constant level of PP over time, deterioration would require some new investment in PP each period. The preference function already discussed implies that the CIG manager is willing to give up some EQ to obtain some amount of PP. Such a preference function is represented by the indifference curve I_c in figure 1, Panel C. The reader may be inclined to argue that PP and EQ are complements rather than substitutes, since causing greater EQ implies that the CIG manager has more PP, but this is not the correct perspective. Clearly the manager would be better off if he or she had more of both goods. What is asserted here is that the CIG manager would be willing to give up some EQ_t for some PP_t. The apparent complementarity is really a joint product and falls properly within the constraints faced by the CIG manager.

During any period, the CIG manager has available various inputs. These inputs can be combined in various ways to produce intermediate outputs which affect EQ and PP in various ways. If the CIG manager were to allocate all of his resources to the production of EQ without producing any PP, he could produce the level R in figure 1, Panel C, *ceteris paribus.* As the CIG manager starts to produce PP, his effectiveness at producing EQ may go up due to the joint product-effect just discussed. However, after some point, additional PP can only be obtained at the expense of EQ. Consequently, the trade-off frontier will look like PR.

The productivity of the CIG expenditures will be increased as the control agency budget is increased. This is because agency implementation activities are complementary to CIG activities. Both the CIG manager and the CIG membership will perceive that the marginal benefits of additional improvements n EQ will decline as more EQ is produced by agency implementation and/or CIG activity. The implication is that the marginal benefits to the CIG manager from an increase in the control agency budget ($MBBI_C$) will decline at higher levels of EQ. Such a curve is shown in figurc 1, Panel H.

The marginal cost to the CIG manager of obtaining increases in the agency budget ($MCBI_C$) increase as TB increases as shown in figure 1, Panel H. There are at least two reasons for this. One reason is that the opportunity costs of other methods of pursuing the CIG manager's goals (such as court action) increase as the CIG manager spends more of a fixed budget on attempting to obtain a higher agency budget. The second reason is that others will be more resistant to budget increase for the agency as the agency gets more budget. Therefore, the productivity of resources spent by the CIG manager for this purpose will decline. The intersection between these two agency budget curves would produce the CIG manager's utility maximizing agency budget. Again, this budget is finite.

Equilibrium in Implementation

Using the elements developed in the model, a steady-state, multiperiod equilibrium can be shown to exist. We will discuss this equilibrium explicitly in terms of the agency budget. Keep in mind, however, that an additional set of conditions is required for court action. While court action equilibrium is not developed here, it is easily shown by analogy. The budget equilibrium exists when desired EQ and TB are the same for all three groups. This will be the case at EQ* and TB* in figure 1.

Recall that all three groups have an interest in the agency budget. Two would like to see it increased, while one would like to see it reduced. Recall also that each group faces costs in creating effective political demand on the legislature. These elements allow for an agency budget equilibrium and an associated environmental quality equilibrium.

In figure 1, Panels F, G, and H show agency budget equilibriums for each of the groups, given a level of effort at effecting the agency budget by each of the other groups. The total demand for the agency budget is the sum, at any budget level, of the effective demands of the three groups. As the budget level increases toward TB*, the agency manager and the CIG manager have higher benefits than costs of affecting increases and will be active in affecting those increases. The firm manager, on the other hand, finds that, at low agency budget levels, the cost of effectively reducing the budget exceeds the benefits; so the firm manager will remain inactive except for gathering information and monitoring activity. As TB* is reached, the firm manager finds it advantageous to expend resources to reduce the budget. At the same time, the agency manager and the CIG manager find that the cost of obtaining additional agency budget outweighs the benefits.

The combination of the agency budget of the three groups (the net demand, D_N) is shown as a downward sloping function on figure 1, Panel D. The horizontal axis represents the total agency budget. The vertical axis represents the net willingness to spend resources to increase the budget which represents the net of $MBBI_A$, $MBBR_F$, and $MBBI_C$. In order to obtain this budget, the environmental demand for government funds must offset demands from other groups and the demand for tax reductions. In order to obtain more funds, the group must expend resources. The amount of resource-expenditure necessary to obtain more funds increases as the agency budget increases. This function can be thought of as the supply of public funds (S in figure 1, Panel D). The equilibrium total budget TB* must be consistent with Panels D, F, G and H.

There is a second set of necessary equilibria. The equilibrium budget can be used, given the laws and court rulings, to produce an equilibrium environmental quality EQ*. This is shown in figure 1, Panel E. The agency will expend resources on DA and EQ according to $\hat{MCI}$ so that EQ* is produced at the marginal cost A. The agency manager will be in equilibrium at EQ* and DA*. This is true for two reasons. First, $\hat{MCI}$ is derived from EP. Secondly, TB* for the agency manager is derived from the benefits and costs of acting on the agency budget developed from the preference and production functions embodied in I_A and DM. The firm manager will be in equilibrium at EQ* and π*. At the budget TB*, the trade-off surface the firm manager faces in figure 1, Panel B, is π_NN which generates this equilibrium. Furthermore, the firm manager is in agency budget equilibrium since $MBBR_F$ is derived at the intersection point in figure 1, Panel G, from π_NN and TB* generates EQ* through the agency. In a similar fashion, the CIG manager is in equilibrium at PP* and EQ*. This is true since the trade-off surface PR is derived from an agency budget of TB* and $MBBI_C$ at TB* is derived from PR and I_C.

Implicit in this equilibrium is an equilibrium in the court action dimension. Subsumed into DM, and consequently MCI, is the productivity and cost of court action by the agency and the cost of responding to court action initiated against the agency by the other groups. Court action by others can affect the shape of DM and MCI, by making it more or less expensive for the agency to obtain greater EQ. A case brought against the agency by the CIG manager is likely to require more control action for a given budget and, thus, shift MCI to the right. Court action by the firm manager would have the opposite effect.

The budget dimension curves of all three groups are also affected by court action. For example, part of the costs reflected in $MCBI_C$ is the opportunity cost of obtaining the CIG manager's goals through court action. Being in budget equilibrium at TB* implies that the CIG manager is also in court action equilibrium. The same is true for all three groups. Thus, the equilibrium depicted in figure 1 representes an equilibrium in EQ, TB, and court action.

APPLYING THE MODEL TO EFFLUENT CHARGES

The model of the implementation process is generally applicable to alternative institutional structures. Suppose that the United States were to adopt an effluent scheme to control emissions. The perspective given by the model helps us analyse the effects of such a change.

But even before we analyze the change, we can analyze the political economy of attempting to adopt the change. Suppose, for the sake of argument, that the effluent change would generate a more efficient outcome than the regulatory system it is to replace. There are gains to some members of society that outweigh the losses to others. A pure effluent-charge system assumes that recipients own the property rights to the environment. Since the current regulatory approach shares property rights between recipients and emitters, it follows that emitters will lose wealth as a result of the change. It will pay for them to act politically through the emitters' lobby to resist the change.

The effluent-charge system appears to be much less complicated to administer than the existing regulatory approach. This might be true if it were adopted in a very simple form, such as the sulphur tax proposed in the early 1970s, which consisted of a flat rate per pound of sulphur emitted anywhere in the US. Since government control agencies would have less to do under the effluent charge and, therefore, less budget, power and discretionary activity, we can expect that control-agency bureaucrats will act politically to resist the effluent charge.

Clearly, it follows from our assumptions that a simple effluent charge will be favoured by the recipients' lobby and opposed by the emitters' lobby and the control-agency bureaucrats. Some sort of compromise will be necessary to obtain adoption. In order to reduce resistance from the emitters' lobby, the effluent charge can be changed into a charge/subsidy system, where emitters are charged if they release more pollutant than current regulations allow and are subsidized by the same amount per unit if they emit less (Mumy, 1980). In essence, the recipient and the emitter would be sharing property rights in the same proportions as they are under the regulatory system. This reduces emitter-resistance but also reduces recipient-support.

In order to reduce bureaucratic resistance, the charge system would have to be modified. One way to do this is to make the charge more complex. The charge could be made to vary from location to location. Variation allows the bureaucrat more discretion and requires more budget. In addition, the bureaucrat will have to enforce the charge. This will require detailed technical information, thus further increasing the bureaucrat's ability to argue for more budget and discretion. All of these factors will tend to reduce bureaucrat-resistance to the change.

These compromises in the design and implementation of an effluent charge could potentially allow the adoption of the system. Alternatively, the simple system could be adopted if the recipients, who are the gainers, were willing to compensate the emitters and the bureaucrats, who are the losers. The political compromises suggested here are, in effect, 'in-kind' compensation. Economic theory indicates that in-kind compensation may be less efficient and can be no more efficient than direct payment of money. This suggests that the political compromises outlined will leave emitters, bureaucrats, and/or recipients less well off than if direct monetary compensation were possible.

Let us suppose that the more complex political compromise is adopted and property rights are shared. How will this affect the implementation process? Refering again to figure 1, the MCI for the agency will be changed. If it is true that the complex effluent-charge system is less costly to administer than the current regulatory system, MCI will shift to the right. This implies that for any budget level (TB), the agency can now produce more EQ (to M′ in figure 2). In Panel A, this shifts DM to DM′ since spending all the agency's budget on DA will not increase the amount available, given that TB* remains the agency budget. This shifts the expansion path to EP′.

The budget constraint for the firm shifts as a result of the change in institutions. Starting from EQ*, the equilibrium environmental quality for the regulatory system, increases in EQ are now less expensive for the typical firm because of the subsidy for the additional control. At the same time, reductions in control would reduce profits because the savings in control costs would be offset by an increase in effluent charges paid. This implies that the profit constraint of the typical firm is shifted to $\pi_{N'}N'$. The constraints facing the CIG manager will also be shifted out (to PR′). This is because the CIG manager is able to obtain more EQ for a given expenditure as a result of the shift to MCI′ of agency costs.

The result of these three shifts is a desire for more TB as well. The institutional change would put each manager on a higher indifference curve. Each manager may have more or less of the other item in his/her goal function (DA, π, or PP), depending upon the exact form of the shift.

These shifts also affect each group's agency budget activity. The firm manager will find it less advantageous to resist agency budget because at least some of the budget may be used to subsidize emission reductions. In fact, if the new equilibrium EQ is indeed higher than EQ* (the regulated equilibrium), the average firm will be receiving a subsidy. This shifts $MBBR_F$ to the right. With less resistance from firms to budget increases, $MCBI_A$ and $MCBI_C$ will also shift to the right. Therefore, D_N (panel D) will shift to the right. The willingness to supply additional budget (the opportunity cost of public funds) will not be affected and it will constrain the budget increase. The net effect of an institutional shift to the complex effluent-charge outlined here would be an increase in the equilibrium EQ and an increase in the equilibrium TB, provided that our assumptions about administrative costs are correct. This result is with conventional wisdom in economics that effluent charges would lead to more efficient results and higher EQ than the existing regulatory approach yields. However, if MCI is higher under a practical effluent-charge system, as may be the case, the equilibrium EQ may, in fact, be lower than the current regulatory system.

SUMMARY

The adoption and implementation of pollution laws in the United States is the result of a complex bargaining process. Of central importance in this process is the issue of who owns the rights to the environment and, therefore, who must pay for controls. The politics and economics of environmental quality in the US have led to a shared ownership between emitters and recipients, much like that in other countries. While it may appear to be the case that recipients have more rights under US laws as compared to other countries, in practice they have less.

A change from the current regulatory system must take into consideration the current distribution of rights. Thus, we suggest that a pure effluent-charge, which gives all environmental rights to recipients, will not be politically feasible. Instead, a compromise will be required such as the complex effluent-charge outlined in the last section. I am not suggesting that this alternative could be adopted or that it is better than other alternatives; only that an analysis of this sort demonstrates some of the necessary conditions for an effective proposal.

Figure 2: Static Equilibrium Model of Pollution Control Implementation.

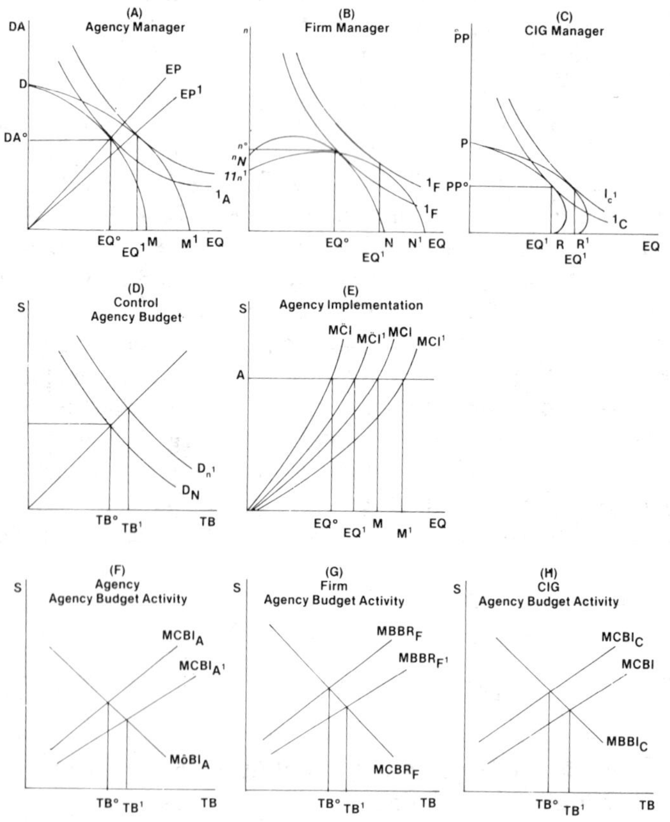

References

Bardach, E. (1977). *The Implementation Game.* The MIT Press, Cambridge, Mass., p. 50.

Becker, G. (1976). *Comment.* Journal of Law and Economics, No. 19, p.245.

Bonine, J. (1975). *The Evolution of Technology-Forcing in the Clean Air Act.* Economic Reporter, at 8 Mono. 21, 25 July 1975.

Downing, P. (1981). *A Political Economy Model of Implementing Pollution Laws.* Journal of Environmental Economics and Management, Fall, 1981.

Downing, P. and Hanf, K. (Eds.) (1981). *Implementing Pollution Laws: International Comparisons.* Volume I, Policy Sciences Programme, Florida State University. See contributions by: Brady, G. and Bower, B., *The Effectiveness of Implementation Incentives in Air Quality Management in the United States;* Downing, P. and Kimball, J., *Enforcing Pollution Control Regulations in the United States;* Ullmann, A., *The Implementation of Air Pollution Control in German Industry.*

Downs, A. (1957). *An Economic Theory of Democracy.* Harper and Row, New York. p. 27.

Downs, A. (1967). *Inside Bureaucracy.* Little, Brown and Co., Boston, p. 84.

Freeman M. and Haveman R. (1972). *Clean Rhetoric and Dirty Water.* The Public Interest, No. 28, p. 51.

Frey, B. and Lau, L. (1968). *Towards a Mathematical Model of Government Behaviour.* Zeitschrift fur Nationalokonomie, No. 28, p. 355.

Goetz, C. and Brady, G. (1976). *Environmental Policy Formation and the Tax Treatment of Citizen Interest Groups.* Law and Contemporary Problems, No. 34, p. 21.

Mumy, G. (1980). *Long-run Efficiency and Property Rights Sharing for Pollution Control.* Public Choice, No. 35, pp. 59-74.

Niskanen, W. (1971). *Bureaucracy and Representative Government.* Aldine, Chicago 1971.

Niskanen, W. (1975). *Bureaucrats and Politicians.* Journal of Law and Economics, No. 18, p. 617.

Peltzman, S. (1976). *Toward a more General Theory of Regulation.* Journal of Law and Economics, No. 19, p. 211.

Quarles, J. (1976). *Cleaning Up America.* Houghton-Mifflin, p. 79.

Sabatier, P. (1975). *Social Movement and Regulatory Agencies: Toward a more adequate—and less pessimistic—theory of Clientele Capture.* Policy Sciences, No. 6, p. 301.

Tullock, G. (1967) *Toward a Mathematics of Politics.* University of Michigan Press, Ann Artor, p. 122.

Williamson, O. (1964). *Managerial Objective in a Theory of the Firm.* Prentice-Hall, Englewood Cliffs.

US Senate (1972). *Statement of Frank Partee.* Oversight Hearings, Committee on Public Works, 16 February 1972, p. 168.

CHAPTER 8

The Integrated Planning System and Economic Reform in Poland

Bohdan Gruchman

PAST EXPERIENCE IN ESTABLISHING AN INTEGRATED PLANNING SYSTEM IN POLAND

The Planning System

THE current Polish planning system, covering physical, socio-economic and environmental aspects is the result of a long evolution dating back to the 1930s.[1] The present shape of the system was mainly influenced by the 1961 law which linked physical planning to long-term socio-economic planning. This system, called 'spatial planning', has been in operation until the present time, undergoing relatively few changes over the years, with the exception of one major change, introduced in 1975 at the regional level, in connection with the administrative reform of the country.

The key element of the entire planning system, as it evolved during the 1960s and '70s, is the National Spatial Plan, which deals with all major physical, socio-economic and environmental problems, thus constituting an integral part of the long-term national development plan. The latest version of the plan, for the period 1971-1990, was finalized in 1975 and contained the following major parts (Grabowiecki and Zawadzki, 1978, pp. 131-143): aims of national spatial development; the settlement system of the country; changes of population distribution and employment; changes in the spatial structure of industry, agriculture and food processing; development of touristic regions; development of technical infrastructure; environmental protection and development including water availability; distribution of social infrastructure and housing construction; reduction of differences in the regional level of living; main development issues of particular macro-regions; instruments of spatial policy implementation; issues of international co-ordination; and directions of further research.

At the provincial *(voivodeship)* level, the national spatial plan had, until 1975, its counterpart in a similarly integrated, general regional development plan which contained the main directions of socio-economic, physical and environmental development for the then-existing seventeen large provinces. Besides this general regional plan, detailed regional plans were prepared for particular parts of the province or development areas crossing provincial

[1]For a brief summary of the early experiences in this field, see Zawadzki (1978, pp. 13-21).

boundaries, where a considerable concentration of investment activities was planned. These detailed regional plans usually covered a smaller area and had a much more elaborate physical planning content, in comparison with the general regional development plan.

With the introduction of forty-nine small provinces in 1975, the integrated regional plans have been replaced by provincial spatial plans with a strong physical planning input. In order to better link such plans with the national spatial plan, an intermediate planning level (macro-planning) has been put into operation since 1975. The country was divided into eight macro-regions and special macro-planning offices of the Central Planning Commission were installed in them. Macro-planning is considered to be part of the national, spatial-planning activity, similarly taking care of physical, socio-economic and environmental issues of development in individual macro-regions of the country. Its particular task is to deal with those problems which are common to several provinces, such as major transportation links, interregional water supply system, professional training above elementary level and so on.

This system of integrated national, macro-regional and provincial spatial plans has constituted a framework for physical planning at the local level—urban plans and physical plans of communes (the lowest administrative level in rural areas). Simultaneously, the integrated national and provincial spatial plans, as elements of the long-term development planning, have been assigned the role of medium-term (5 year) socio-economic planning at the national, provincial and local levels. Annual socio-economic planning, in turn, was supposed to be an implementational instrument of the medium-term planning.

Operation of the Planning System

The Polish planning system represents an integration of long-term physical, socio-economic and environmental planning at the national and regional levels, while leaving as separate issues physical plans at local level and medium- and short-term socio-economic plans at all levels. Such a compromise solution has many advantages: it helps to deal with major development issues simultaneously from all relevant points of view and, at the same time, leaves room for specific problems and approaches, represented by each planning activity in areas where this specificity is particularly strong — at the local level (physical planning) and in the medium- and short-term horizons (socio-economic planning). The successful operation of such a planning system depends on a number of conditions, two being of particular significance. Firstly, the integration of physical, socio-economic and environmental planning should not only be formal but also have substance, through concepts and methods. Secondly, effective links are necessary between the integrated plans and those issues dealt with separately.

The evaluation of the operation of the planning system has been highly variable; some experts, while pointing to its shortcomings, have expressed a qualified positive opinion about the system; others, particularly lately, have criticized the system, claiming it to be totally ineffective (Piatek, 1980; Regulski, 1981). The question of the system's adequacy to deal with the new conditions which will be created by the forthcoming economic reform in Poland will be discussed later. But, under the circumstances which prevailed in the country up until 1980, one can point to several positive and negative aspects of the planning system's operation.

Despite shortcomings in details, much valuable experience was gained in the preparation and casting of the national spatial plan for the 1971-1990 period. After several futile attempts at preparing such a plan in the late 1950s and '60s, a plan was drawn-up in the early 1970s which constituted a truly integrated planning document where physical, socio-economic and environmental considerations were simultaneously taken into account. What is particularly noteworthy in this plan is the inclusion of the results of many environmental and ecological studies, for example 'active' ecological studies on particular socio-economic developments and 'passive' studies on the impact of economic growth on the natural environment.

At the regional level, the preparation of detailed regional plans, for areas with a particular concentration of investment activities, played a useful role in the guiding development and shaping other kinds of planning. These regional plans were genuinely integrated plans with a concrete socio-economic content, spelled out in physical and environmental terms; they also provided a good basis for drawing up local physical plans (especially urban plans) and making decisions on investment and location. Such plans were prepared for several industrial zones, such as the Konin brown-coal region and the Legnica copper zone, and for major urban agglomerations like Warsaw, Gdánsk and Cracow. The conditions which made the detailed regional plans a successful exercise in integrated planning did not exist before 1975, when general regional plans were elaborated for all 17 *voivodeships* (provinces). These general plans were vague on the socio-economic side and weak in their physical and environmental aspects.

The shortcomings of the general regional plans are, to a considerable extent, part of a more general weakness in the entire planning system as it operated until recently. There were insufficient links between the national spatial plans and their provincial counterparts—until 1975, as we have seen, with the general regional plans and, since then, with the provincial spatial plans for the smaller *voivodeships*. The socio-economic content of the latter plans (in their first generation, prepared after the 1975 reform) was particularly weak, owing to insufficient connection with the national spatial development plans. Thus, in practice, the provincial spatial plans remained only physical in character, with a very limited socio-economic content. Their impact, consequently, on the local physical plans was reduced and, even more so, their influence on the medium-term socio-economic planning at the provincial level.

When stressing the weak links between various types and levels of planning, it is not so much a formal interdependence that is needed between them (as some central planners consider the most important aspect), but rather an appropriate correspondence in content, planning horizon, concepts, co-ordination criteria and methods. These have been either lacking or insufficient between the national and provincial spatial plans. The solution was sought in the introduction of the macro-regional planning level and the establishment of special agencies to do the job. Unfortunately, in the first five years of their existence they have not, on balance, produced the expected results.

Another major, if not the main, shortcoming of the planning system has been its weak implementational power. Naturally, considering the type of planning we are dealing with here, this implementational power must be understood, in a broad and rather indirect sense, as an impact on operational types of planning and on various instruments of implementation. According to this line of reasoning, the relatively weak links have already been mentioned between the integrated plans at the national and provincial level, and physical plans at the local level, which through zoning regulations, locational permits, etc., have the legal power to regulate economic activities. In addition, physical plans have had their own defects—a static character, short-lived validity and possible manipulation by implementational agencies.

On the other hand, the long-term national and regional spatial plans did not exert a strong influence on the 5-year and annual socio-economic plans. The main reason for this situation lay in growing voluntarism of central decision-makers who, particularly in the second half of the 1970s, disregarded the already approved long- and mid-term plans and made investment decisions often contrary to those already in the plans, thus very soon rendering them obsolete. Such practices gradually became the rule and, as a consequence, a general crisis concerning types of longer-term planning was the result. These included integrated forms of planning at the national and provincial levels. This crisis has grown even deeper, owing to the very limited range of instruments through which the long-term spatial plans (particularly at the national level) could more directly influence implementational decisions. Such instruments, with spatially differentiated impact, have not yet been sufficiently developed, despite studies underlining their usefulness, even in a centrally planned system (Gruchman, 1967). The common belief was that the plans themselves are sufficiently effective instruments of decision-making and implementation with which the centre could steer the development of nationalized enterprises.

In consequence, the only implementational instrument of long-term planning worth speaking about was the system of locational decisions. Location of larger investment projects required a decision from the Ministerial Council or, at least, from the Central Planning Commission. Provincial authorities co-operated in the locational proceedings at the centre and were solely responsible for decisions concerning smaller investment projects. All locational authorities were required to base their locational selections on the national and regional spatial plans as far as the general location was concerned (such as the determination of a particular settlement) and on the physical local plans for the detailed location (i.e. referring to the site within a settlement). This locational system did not function properly. There were gaps in the locational procedure, for example major expansions of existing facilities were omitted. Many plans were too vague to justify certain locational decisions and sectorial authorities exerted effective pressures in order to secure the most convenient location. In addition to all this, the previously mentioned voluntarism of key decision-makers in the country resulted in a number of faulty place-selections which disregarded locational rules inscribed in the relevant plans.

THE PROPOSED ECONOMIC REFORM AND ITS CONSEQUENCES FOR THE PLANNING SYSTEM

Main Features of the Economic Reform

The main feature of the economic reform currently under preparation is the transformation of a centrally and, in recent years, very tightly managed national economy into a decentralized system in which the key role will be played by independent and self-managed enterprises (Trybuna Ludu, 1981 a). The direct orders and detailed normative guidelines from central authorities, which did not allow enterprises much freedom of action in the past, will be replaced by indirect instruments of managing the national economy. Among the latter, predominant roles will be assigned to a proper price system, tax and credit policy, various financial norms and legal instruments. An expanded autonomy is also being proposed for the communes and provinces in solving their local problems. To this end, the scope of the local economy, sharply reduced as a result of the 1975 administrative reform, will again be enlarged and the financial base of the local authorities substantially strengthened. At the central level, in turn, the vast bureaucratic apparatus necessary to manage the day-to-day affairs of the national economy will become superfluous. Therefore, the number of industrial ministries will be reduced and the numerous industrial boards ('watch-dogs' of all enterprises belonging to a particular branch) will be abolished and, where necessary, replaced by voluntary associations of independent enterprises.

Naturally, in a new economic environment, the overall planning system itself has to undergo substantial transformations. It can no longer exist as a hierarchical system with the national plan, directly and in all aspects, subjugating all other plans which are dependent on it. The bulk of enterprises will, independently, prepare plans of their own with the exception of some key enterprises producing basic commodities for the entire economy. These will be required to prepare 5-year development plans. The obligatory planning directives and norms, which tied the lower-level plans to the central national plan, will now be replaced by information flows going in both directions with a simultaneous preparation of plans on all levels. Central planning will thus lose its operational nature and change into strategic planning for the structural transformation of the economy.

In accordance with these new factors, annual socio-economic planning, which was the main guiding instrument of the national economy in the past, will lose its importance. Instead, central planning for five-year periods will be upgraded as the main planning activity of the central government. Through this process will be created the framework for development and operation of enterprises, including decisions about direct government-investment activity. Such a 5-year governmental plan will only work if it is soundly based on a long-term development scheme. Hence, long-term planning activities will play a much more crucial role in the future than they did in the past.

With the central government development plans losing their obligatory character for enterprises and local authorities, the main question in the new situation will be how to translate the decisions of structural transformations, inscribed in the long- and medium-term plans, into effective instruments of implementation. Another problem of paramount significance is how to secure substantial public participation in the determination of developmental aims and in control of the planning process and its execution. Hitherto, the role of the public in these fields was rather formal and superficial. But discussion of these two issues (effective implementation and public participation) will follow in more detail later.

Proposals for the Integrated Physical, Socio-economic and Environmental Planning System

The outlines of the economic reform, published by the Central Commission of the Economic Reform, deal with integrated forms of planning in rather general and incomplete ways. More specific are several proposals formulated by various specialized committees, professional organizations and individual experts (Trybuna Ludu, 1981 b; Gorzelak *et al,* 1981). What emerges from all these proposals is a desire to establish an integrated planning system which would be adequate for the new situation created by the economic reform, while simultaneously profiting from the past experience of spatial planning in Poland.

All the proposals agree on the central role to be played in the new system by a long-term national development plan, broken down by regions and macro-regions. The more detailed characteristics of the plan's content leave no doubt about its truly integrated nature. However, some proposals differ about the time horizon of such an integrated plan. The Central Committee's project recognizes the fact that the physical and environmental aspects of the integrated plan may require a time horizon longer than the 10- to 15-year span suggested for the socio-economic content. What is even more significant is the proposal that, in the pre-planning period (i.e. during the phase preceeding plan preparation) among other studies, a physical development programme should be worked out reaching far into the future.[2] This programme would create a spatial framework within which, later on, directions of socio-economic development could be outlined in the long-term integrated plan. Other proposals do not emphasize the time horizon of the integrated national plan, stressing only its long-term nature.

All projects on the future planning system stress the need to have an integrated plan at the regional level. Some authors even consider the regional level, and not the central one, to be the most crucial element of integrated planning. It is on that level, they argue, that physical, socio-economic and environmental aspects have a common spatial dimension which greatly facilitates integration. There is, however, a difference of opinion as to whether the term 'region' should stand for the administrative provinces or for some other areas. One proposal is firm in considering plans for the provinces as the only form of genuinely integrated plans. Planning for other spatial units is assigned only a supportive role and does not necessarily have to be integrated.

Other proposals envisage regional plans not only for provinces, but also for special development regions (such as those with a concentration of investment) and urban and industrial agglomerations (metropolitan regions). Macro-regions, as units of integrated planning, are assigned two different functions: either as main spatial units in the integrated national plan, or as voluntary associations of provinces with the aim of mutually solving development problems of a common nature. It seems that both functions could be accommodated as they serve different useful purposes.

The proposals are divided on the need for integrated long-term planning at the local level. Some suggest that the integrated system should reach that level as well; integrated long-term plans at the local level would be starting points for any other kind of plans for cities and communes. Others recognize the difficulties connected with the preparation of such elaborate plans at this level, including the difficulty of determining meaningful, long-term socio-economic development from the local point of view. Therefore, such proposals accept the past experience of separate long-term physical and medium-term socio-economic plans, requiring them to be closely linked with each other.

The desire to operate with integrated plans manifests itself also in medium-term planning. If we accept the function of such planning to be an implementational device for structural changes embodied in the long-term plans, then consequently we should go one step further and see to it that medium-term planning reflects all the dimensions of planning for a wider horizon. Proponents of such a development envisage integrated medium-term

[2]Such a perspective programme would somewhat resemble the General scheme of development and distribution of productive forces, elaborated in the late 1970s in the Soviet Union (Nekrasov, 1978).

plans both at the national and regional levels. Others do not go that far; in view of the future local governments' autonomy, 5-year plans at this level should only cover enterprises and institutions which are owned or directly supervised by local authorities.

SPECIAL ISSUES OF INTEGRATED PLANNING

Public Participation in Integrated Planning

Public participation in integrated planning is not only a matter of democratic principles, but also an issue of particular importance when dealing with such a complex and multidisciplinary area as that covered by integrated planning. Determination of social, economic and environmental goals, setting priorities and resolving conflicts among them, formulation of development criteria with which to measure progress and many other similar issues cannot be left solely to the administration.

The planning system that existed in Poland until recently had a number of devices built-in for the representatives of the public to be involved in the planning process. Five-year and annual plans were passed by Parliament and the integrated long-term plan was consulted by relevant commissions of Parliament. Before being adopted by the Government, the long-term plan had to be accepted by the National Council for Spatial Development (a special governmental body attached to the Planning Commission) which included representatives of central and provincial government, scholars and experts from various professional and civic organizations. Bodies with a similar membership also operated at the macro-regional level. A considerable research input in the integrated plan preparation has come from the Committee for Space Economy and Regional Planning, a scientific organization attached to the Polish Academy of Sciences.

At the provincial level, spatial plans had to be accepted by People's Councils, consisting of members elected by inhabitants of the province. A similar procedure was prescribed for medium-term and annual socio-economic plans. In turn, physical plans at the local level had to be presented to the relevant communities for their comments before approval was obtained by the local People's Council.

Despite all this, public participation in planning has not been of great significance. Many diverse factors have contributed to this state of affairs, some of them playing a major role. Thus, representatives of the public did not have much freedom of choice in a highly centralized planning system. This was particularly evident at the provincial and local levels. The planning authorities submitted plans for approval in the final stage of their elaboration, with no alternative proposals, so that there was no time to introduce substantial changes into the planning document. Finally, the selection process of public representatives had many shortcomings as well as the rapport of the elected persons to their constituencies.

The new planning system is being devised in such a way as to remove all such obstacles to public participation in planning and to shield it from voluntarism of individual decision-makers. As previously mentioned, public participation in planning is particularly crucial in spelling out societal goals of integrated development. It is obvious that, in such a process, the economic growth goal (dominant over all others in the past) will be adjusted to proper proportions in relation to social and environmental issues. In order to secure their strict adherence in the planning process, it is proposed that planners should submit their work to public scrutiny at least three times: at the initial stage, when goals are set and translated into alternative planning targets; next, when the plan has been drawn-up in detail; and finally when any changes need to be introduced to the plan during its execution.

The requirement to submit for approval an outline of development goals and planning targets is a safeguard against the formal acceptance by planners of overall societal goals, which later on are twisted in the process of spelling out planning targets. It was at this stage in the past that many environmental goals were eliminated or substantially reduced in importance in favour of economic goals. When the blueprint of the plan is submitted for approval, the representatives of the public should see to it that the most favourable development alternatives have been selected and that no departure from the earlier determined goals and targets have taken place. Public approval of changes in the plan during its execution is necessary to prevent government agencies substantially altering the implementational course—a practice commonly followed in the past. Apart from these devices, central and local government agencies will be evaluated on their performance when annual reports about plan-execution are submitted for approval by Parliament and People's Councils.

Naturally, this discussion cannot cover all forms of public participation in integrated planning which also involve the media, professional organizations and civic pressure groups.[3] Each has a role to play in order for public participation in planning to be widespread and effective.

Instruments of plan implementation

Formerly, integrated planning had only a limited possibility of directly influencing the economy and society at large. It did so with mixed results in an indirect fashion through operational forms of planning which used to be linked with the old system. The new system of planning does away with the formal and hierarchical dependence of enterprises and local governments on central government plans. Hence, the search for effective implementational

[3]Pressure groups in this area are active in many parts of Poland. In the autumn of 1981, the first Congress of the Polish Ecology Club took place in Cracow, in order to work out a detailed programme of popular action for the protection of the environment.

instruments, tied to the new planning system, becomes of great importance.

The existing locational procedure must be adapted to the new situation and considerably strengthened. It will be able to play its previous normative role, but only in the markedly reduced sphere of investment activities, directly organized by the central government, and in investment projects of local governments, mainly outside the material production sectors. However, the bulk of investment activities in production will be organized by independent enterprises, which will not have to obey location proposals formulated by central and local governments. The latter can only legally protect certain areas, such as parks, water catchment zones, historic sites, etc., from particular activities. To this end, the legal force of locational procedure will have to be strengthened, becoming more closely allied to the physical and spatial planning system.

In order to influence the locational decisions of enterprises in a more positive and effective way, the directives of economic reform foresee the introduction of two economic instruments, namely, investment credits with different rates of interest and locational fees which enterprises will have to pay to local governments for the construction of infrastructural overhead facilities. This is a most welcome development, provided that both measures operate in accord with the integrated planning system. However, it is doubtful whether they will suffice as parameters influencing economic activities. If enterprises accumulate enough capital on their own, they will not need to resort to bank credits to implement a construction project. On the other hand, not all projects will require investment in local infrastructure. Therefore, it seems advisable to search for more implementational instruments which could have a differentiated impact in space.

Three such instruments to be used for influencing enterprises are envisaged by the economic reform, namely, a tax on payroll or number of employees working in enterprises, a tax on area and quality of land occupied by them and finally, a tax on water and mineral resources used by enterprises. In favour of differentiating, the first tax concerns the regionally conditioned supply and demand for labour. The second could be utilized as a tool for directing enterprises away from overcrowded agglomerations, with a more rational use of the land in the areas. Finally, taxing water consumption and utilization of mineral resources, in a spatially differentiated fashion, could help to preserve them and, simultaneously, to redistribute relevant economic activities in a desired manner. The last tax could be expanded to also include environmental pollution or perhaps a separate tax should do this job.

The list of possible devices similar to these could be much longer, particularly if local taxes and other economic measures are taken into account. The crucial issue concerning their application is how should their values be determined in order for them to serve societal development goals and be fully incorporated in integrated development plans. Their determination by market forces cannot, for the foreseeable future, be taken into account in the circumstances existing in Poland.

The direct calculation of these parameters from integrated plans will, most probably, not be possible for many of them. Therefore, it seems advisable to explore the possibility of applying a system of shadow prices derived from integrated plans, along similar lines as those suggested by social cost-benefit literature for developing countries. (UN, 1972; Little and Mirrlees, 1974; Little and Scott, 1976). Such shadow prices could then help to determine the value of particular parameters actually applied. This is, however, an area needing much more research and experimentation than has hitherto been done in either the East or the West.

References

Gorzelak, G., Kolipiński, J. and Kulesza, M. (1981). *W sprawie uwzglednienia zagadnien przestrzennych w reformie.* (Remarks concerning the inclusion of spatial problems in the Reform). A paper prepared for the Committee for Space Economy and Regional Development, Warsaw, p. 20.

Grabowiecki, R. and Zawadzki, S. (1978). Metoda opracowania planu przestrzennego zagospodarowania kraju do 1990r. (The method of preparation of the National Spatial Plan until the year 1990). In *40 lat planowania struktury przestrzennej Polski.* (Forty years of experience in planning the spatial structure of Poland). B. Malisz (Ed.). Polska Akademia Nauk, KPZK, Studia tom LXIV, PWN Warsaw, pp. 131-143.

Gruchman, B. (1967). *Czynniki aglomeracji i deglomeracji przemyslu w gospodarce socjalistycznej na przykladzie Polski.* (Agglomeration and deglomeration factors of industry in a Socialist Economy. A Case Study of Poland). PAN, KPZK, Studia t. XVIII, PWN Warsaw, p. 151.

Little, I.M.D. and Mirrlees, J.A. (1974). *Project appraisal and planning for Developing Countries.* Heinemann Educational Books, London, p. 488.

Little, I.M.D. and Scott, M.F.G. (1976). *Using Shadow Prices.* Heinemann Educational Books, London, p. 253.

Nekrasov, N.N. (1978). *Regional Economics: Theory, Problems, Methods.* (In Russian). Ekonomika, Moscow, p. 343.

Piatek, S. (Ed.) (1980). *Podstawy prawne i instytucjonalne systemu gospodarki przestrzennej.* (Legal and Institutional Foundations of the Space Economy). PAN Instytut Geografii i Przestrzennego Zagospodarowania, Biuletyn Informacyjny, z. 32, Warsaw, p. 140.

Regulski, J. (1981). *Somorzadne wladze terenowe w zreformowanym ustroju gospodarezym PRL.* (Self-governing local authorities in the Reformed Economic System of the People's Republic of Poland). Konwersatorium "Dóswiadczenie i Przyszlośó". Grupa roboeza: Prawo, praworzadnośé i gwaraneje prawne. Warsaw, p. 33.

Trybuna Ludu (1981a). *Kierumki reformy gospodarezej. Projekt.* Warsaw, p. 128.

Trybuna Ludu (1981b). *Zalozenia do projektu ustawy o radach narodowych.* (Premises of proposals for the law regarding Local People's Councils). Warsaw, p. 16.

United Nations (1972). *United Nations Industrial Development Organization, Vienna: Guidelines for project evaluation.* United Nations, New York, p. 383.

Zawadzki, S. (1978). Od planowania regionalnego do planu krajowego. (From regional planning to the national spatial plan). In *40 lat planowania struktury przestrzenej Polski.* (Forty years of experience in planning the spatial structure of Poland). Polska Akademia Nuak, KPZK, Studia tom LXIV, PWN Warsaw, pp. 13-21.

CHAPTER 9

Experiences of Environment Conservation in theUzbek SSR

(abstract)

Igor V. Chirgadze

THE experiences of social and economic changes in the Uzbek SSR, as well as in other republics of the Soviet East, is of great international importance, particularly valuable for developing countries. Recently, special measures have been formulated for the protection of the environment and the rational use of natural resources under a separate economic administrative section. Prompted by the increase in the scope of economic production, considerable expansion of work was required concerning environmental protection and the rational use of natural resources, with a consequent strengthening of its programme orientation and the complex infrastructure all this implies. The principal objective has become the improved co-ordination of activities by the authorities in different branches of industry and also the various ministries and government departments involved.

Accordingly, in the tenth five-year Plan for the economic and social development of the Uzbek SSR, (as well as in other Soviet Union Republics), a special section has been introduced with comprehensive measures relating to environmental protection and the rational use of natural resources. In this section of the Plan, environmental protection objectives are set before ministries, government departments and territorial authorities (the Executive Committees of local Soviets of People's Deputies). The wise use and protection of water and atmospheric resources, as well as land, forest and mineral resources, are the main components of the outlined measures. Special attention has been given to the development of reservations and national parks, including protection of wild animals, birds and fishstocks. More than 350 million roubles have been allocated for these purposes, half of which goes towards the protection and rational use of water resources

With the rapid growth of industry, transport and agriculture, and the consequent use of still greater amounts of natural resources, protection of the environment and the rational use of resources have become of prime importance in the economic and social objectives of the Soviet State. This environmental commitment was outlined many years ago in the Constitution of the USSR, which states: "In the interests of the present and future generations, the necessary steps are taken in the USSR to protect and make scientific, rational use of the land and its mineral and water resources, and the plant and animal kingdoms, to preserve the purity of of air and water, ensure reproduction of natural wealth and improve the human environment."

CHAPTER 10

The Effects of Socio-economic Activities on the Environment in the Framework of Integrated Physical Planning: Some French Experiences

Francis Wasserman

INTRODUCTION

UNTIL recently, the processes of economic planning and the geographic assignment of economic activities were handled relatively independently of one another. Under such conditions, it was nearly impossible to include the environmental effects of socio-economic activities within the framework of economic planning. It was considered that the localization of economic activities should be carried out 'spontaneously' under the double influence of 'natural' factors and economic conditions. This situation lasted until the crisis of 1929.

The establishment of the Tennessee Valley Authority in the United States marked an important step in the awareness of the theoretical insufficiency and overly marginalist thought occurring during that period. For the first time, a geographic river basin was considered as a whole, hydraulic development binding together concerns extending from electricity production to irrigation, drinking water supply and the overall development of water use.

Obviously, this did not prevent the subsequent repetition of the same mistakes. In the United States, as well as elswhere, too many productive investments remained designed for, and calculated on the basis of, distant goals and without consideration of the external economic costs brought about by such conceptions, especially since they did not fit adequately into the natural environment. Perhaps it was the mistakes observed in the use of the natural environment, following certain large projects,[1] which slowly but surely led to an emerging awareness of the necessity for a greater overall view of economic and physical development.

Actually, this awareness did not stop with the analysis of large projects and was subsequently present in the consideration of all aspects of development. In this manner, planning, which had previously been conceived solely in terms of the productive sectors, gave way to a more geographical conception of economic development: balanced regional development was to become recognised as a necessity. This concept was supported by a mixture of theoretical and practical considerations, among which the environment must have a place. In order to clarify these considerations, a brief historical survey

[1]Especially dams: The Rhone River Development in France, Aswan Dam in Egypt, etc.

of the French experience follows. It will show how purely economic planning gave way to balanced regional planning with the environment being taken into account to an ever-increasing extent.

THE FRENCH EXPERIENCE IN REGIONAL PLANNING

Guiding Principles

The instinctive reaction of a French government, faced with the constraints of the international division of labour in the decade 1950 to 1960, was to protect declining sectors through a series of measures intended to postpone that decline. These measures included subsidies, sustaining prices, reduction of transportation fares, protective tariffs and so on. They resulted in maintaining considerable disparity between the production levels of certain national and foreign industries. The acknowledgement of this fact gave birth to the idea that a good policy consists of less in maintaining declining activities at whatever cost than in providing recession-affected towns or zones with replacement activities.

A recovery policy was then instituted in a number of sectors (mineral basins, textiles, rural life). This policy, conceived in terms of economic geography, eventually took the place of strictly sectorial policy. In concrete terms, this policy consisted of the desire to maintain French urbanization within 'reasonable' limits and to sustain those areas of provincial growth which could have the effect of bringing about regional development. The idea of those promoting this re-orientation was that, when the necessity for reconversion was accepted, it should more or less consciously take into consideration the external economic costs of an exodus of the population. The practical difficulties involved in creating new activities in recession zones should then lead decision-makers to analyse the problem of external economic costs more acutely than in the past: if industry managers do not localise their investments here, is this not related to a distorted price system? The spontaneous industrial concentration in the Paris region, for instance, can be explained by the lack of direct invoicing of certain equipment costs to business, costs which are in fact supported collectively by the rest of the nation.

It would then appear that such an awareness could only expand from the observations already made on individual development operations to more global, economic levels. In addition, it appeared that the price system businessmen were using to make their decisions was not only distorted in static terms. There were also problems with the dynamic aspects because of the excessive weight of short-term planning in decision-making. It is clear that such a system could only end in explosive cumulative effects, causing a few pockets or axes of development to co-exist with large zones of recession.

Actions Taken

On the basis of these considerations, the balanced regional development policy of the French administration has consisted, since 1963, in the installation of the institutional and economic tools required for action. The recognition of the need to forcibly break the sectorial framework of economic policy led to the need for modifying the framework of institutional action which was being used by the administration. Up to this point, actions had been taken through the channel of sectorial ministries, with the Commissariat Général du Plan (General Commission of the Economic Plan) reinforcing this sectorial character because it lacked the means of giving its objectives a truly geographic scope.

Two organizations were set up:

(a) The Commission Nationale d'Aménagement du Territoire (CNAT) (National Regional Planning Commission), composed of high-ranking public officials, representatives of economic activity, unions and local communities.

(b) The Délégation à l'Aménagement du Territoire (DATAR) (Regional Planning Delegation).

The CNAT was responsible for defining geographical objectives, while DATAR was responsible for the implementation of these objectives, using two means of action:

(a) a capacity for systematic intervention through the regionalization of credits of the various ministries; and

(b) its own endowment, the Fonds d'Intervention pour l'Aménagement du Territoire (Funds for Activities in Regional Planning).

In addition, a policy consisting firstly of deconcentration (attributing greater powers to provincial officials), then decentralization (an increase in power, especially in the financial field, conceded to community then to regional organizations) was begun. This made it possible to set up the basis that would, in turn, open possiblities for actions directed towards integrated planning.

On the national level, action linked to balanced regional development took two main directions:

(a) localization of employment, previously mentioned, as a starting point; and

(b) protection of nature and the environment.

Protection of Nature and the Environment in Regional Planning

Linked to the general concerns of economic-geographic policy, an entire policy for preserving rare sources was developed in 1963, within the framework of balanced regional development. This policy was based on the desire to reduce certain external aspects, essentially concerning the management of large natural areas (regional and national parks and conservation of the littoral zone), national resources (air, water) and pollution and nuisance control (water, air, solid waste, noise). Since its creation in 1970, the ministry responsible for the environment has taken up and continued this policy. Nevertheless, action in balanced regional development has been marked in certain fields with a large geographical component. One of the most remarkable successes has been in the area of water management, with the creation of Agences Financières de Bassin (River Basin Financial Agencies) in 1964. This represented the first coherent attempt to internalize certain external, and previously ignored, effects.

The hydro-geographic basin was used as the basis of a single physical and geographic unit. Based on the physical reality of a hydro-geographic basin, the Basin Agency is the organization that expresses the solidarity (in an ecological sense) between the various water users within the basin. The mission of the Basin Agencies is to aid in the building of installations of public interest and in both the development of water resources and water pollution control. As public establishments with financial autonomy, they are involved in administrative, financial and technical activities.

The agencies establish and receive two sorts of funds: pollution taxes on effluent waste as well as fees for the use of surface and underground water. These are the sources of revenue for the agencies. These revenues are eventually redistributed in various forms:

(a) carrying-out or financing studies;
(b) subsidising low-interest loans for carrying-out work (dams, purification plants, pollution control equipment, flood protection installations, etc.);
(c) bonuses for good operation of water-purification or pollution-control installations; and
(d) personnel training.

OTHER ASPECTS OF INTEGRATED PHYSICAL PLANNING IN FRANCE

So far, discussion has focussed on how regional development and planning in France have evolved from a purely economic point of view to the consideration of environmental problems in relevant geographical units. The present state of physical integrated planning has in addition been determined by the evolution of macro-economic and sectorial planning, as by the introduction of some regulations related with the protection of the environment (such as the obligation of environment impact statements).

The Evolution of Sectorial Planning

Since World War II, France has been known for its so-called 'indicative' planning process used on the macro-economic level. This process has slowly evolved towards both sectorial and regional planning.

The creation, in the late 1960s, of twenty-two economic administrative regions led to the need for some kind of regionalization of the French National Plan, even though these regions lacked any real power. The first steps were taken during the Seventh Plan (1975-1980). About twenty-five priority programmes were designed, with strong financial commitments. One of these took into account the protection of the natural environment. A few others had objectives indirectly linked with environmental policies. An attempt to regionalize these priority programmes was made during the preparation of the same Plan.

Though no real sectorial planning on the regional level, and no real regionalization of the national sectorial planning, were done, these attempts to design thematic priority programmes for the regions can be considered as important steps towards some kind of integrated physical planning approach.

Environmental Impact Assessment

Other means have been envisaged in France to link integrated physical planning and the environment. In 1976, France adopted legislation imposing environmental impact statements. Following the US experience with NEPA (National Environmental Policy Act), this new regulation had, among its main objectives, the spreading of behaviour consistent with integrated physical planning.

As opposed to other environmental regulations, the environmental impact assessment legislation is not a substantive law (no physical standards are set) but it imposes legal proceedings. These are indirect means of acting on the environment by the granting or refusing of various authorizations (certification of chemicals, building permits, development plans, etc.). By making all the information public, it is possible to draw up an overall evaluation of projects in terms of their ecological components.

The environmental impact report, which is a synthesis of this data, should contain explicit information on:

(a) the analysis of the initial condition of the environment and its evolution, if there were no action;
(b) the various impacts of the project, distinguishing those that cannot be avoided;
(c) the alternatives;
(d) the irreversible processes initiated by carrying out the projects and the long-term effects of these processes; and
(e) the mitigation of negative impacts.

In the absence of an environment impact statement or if the study is incomplete, the project can be brought to court and blocked. Over the course of time, this possibility has been used and has resulted in a precise jurisprudence that sets the terms and conditions for evaluating the effects of socio-economic activities on the environment.

About 4,000 environmental impact statements are made every year. About half of these concern industrial projects. The other half deal with urban development projects, rural development projects, transportation infrastructure, and others. Due to the importance of these latter types of project, some links have been established between the impact statement procedure and the various kinds of spatial planning used in France (such as urban planning and zoning and rural planning). This trend results in carrying-out impact studies at the level of overall programmes rather than for particular projects. This empirical form of integrated physical planning is emphasized by other repercussions:

(1) The mobilisation of expert capacities (especially in the field of ecology) and the activity of certain social groups, previously untapped;
(2) A social apprenticeship was served in the attempt to find a consensus between the various participants;
(3) The abandonment of numerous projects situated in sensitive zones; and
(4) The appearance of veritable doctrines of development, following the accumulation of modifications in certain projects brought through these impact studies.

All these consequences have led towards improvement in the quality of decisions made. In effect, the environmental impact statement requires an evaluation of the effects of decision-making processes on the environment and requires a global and multidisciplinary point of view. This 'systems' approach is entirely analogous to the approach that should provide an orientation for all work in integrated physical planning. The parallel development of the practice of environmental impact studies and the methodology of ecological planning reveal the adequacy of the impact study procedure on the one hand and, on the other, the necessities of taking the environment into account in physical planning. This could explain the success of this procedure.

Some Recent Developments

Some new developments in planning have been introduced since the change of government following the May 1981 elections. The general trend seems to rely more heavily on national and sectorial planning and, simultaneously, to develop regional and local planning. A new legislation has been adopted to establish a real decentralization and regionalization in the decision-making process. Though it is quite impossible at present to know what the real consequences of this legislation will be, it seems that positive impacts may be expected for the development of integrated physical planning.

THE CASE OF WATER POLLUTION

The case of water pollution is one of the most interesting examples of integrated physical planning in France. It has been seen that the Basin Agencies play an important role in co-ordinating various water uses. This is done in relation to a long-term policy, based on quality objectives. These objectives, to be achieved for all water courses and bodies within a given time horizon, provide the basis on which decisions on investments and discharge authorizations can be made. They are stated in terms of 'types of uses possible', implying specific levels for the various pollutants' concentrations.

The quality-objectives procedure consists, firstly, in defining a development plan (demographic, social, economic and spatial aspects) which can be used to determine the various desirable uses of water (drinking water, irrigation, industrial uses, recreational activities or tourism) for the next ten years. The best compromise should be made between possible *a priori* uses and, in this manner, the latent conflicts present in water appropriation within the basin should be resolved. Finally, with the quality objectives set in geographic terms, the levels of quality necessary for the uses in question are evaluated. The definition of these objectives, therefore, makes it possible to compare them with the existing situation, specify actions to be carried out and authorize long-term water use corresponding to a chosen development plan.

The procedure used to establish the maps of water-quality objectives involves the local agencies of all the concerned ministries (environment, agriculture, etc.). The maps are drafted for each sub-basin, so as to give the following information:

(a) an initial assessment of water quality;
(b) the quality objectives; and
(c) the costs to achieve these objectives, taking into account the investment possibilities of the economic activities.

Associations and users are consulted during this drafting phase and alternative maps may be submitted. In the end, a more formal consultation phase begins with local agencies. The final objectives ensure that the desired water-quality costs are acceptable to the various actors. As the quality objectives set the necessary discharge reductions and forecast the corresponding capital and operating costs, the implementation of required actions may be considered immediately.

The quality-objectives policy has several major advantages:

(1) the elaboration of a coherent set of actions;
(2) the consideration of all actors and factors influencing water quality;
(3) the balancing of costs and benefits through public consultation and hearings; and
(4) consistency of the water-quality objectives with the broader objectives of the National Economic Plan.

This procedure of quality objectives is an exemplary case of taking into account the environment in integrated physical planning. In effect, a given development plan permits an expression of the interrelations between socio-economic activities and the state of the natural environment, these two concepts being placed on the same level. In addition, it lends itself easily to public participation. Finally, the historical development of this process bears witness to the progress that can be accomplished in refining the concept of planning by including the environment among the criteria to be taken into account.

CONCLUSIONS

The French case of water pollution, just presented, is exemplary in two respects. Firstly, it shows the changeover from sectorial planning to balanced regional development, a policy that takes the environment into account. And secondly, it illustrates, at the level of the Basin Agencies, the co-ordination between planning policy through prices (fees and taxes) and a policy of integrated physical planning (with quality as the objective), which rests on a procedure of public accord.

The preceding examples show that, from an institutional point of view, it is possible to encourage consideration of the environment in integrated physical planning, even when starting with very different situations. On the other hand, the methods used to initiate this process of taking the environment into account are usually very similar and, for the most part, arise from ecological planning. It appears that contexts and initiatives of very different natures can lead to the setting up of a favourable climate for considering the environment in physical integrated planning. Nevertheless, no satisfactory methodology has clearly come to the fore. If ecological planning, especially when applied within the framework of environmental impact studies, appears to be a particularly interesting instrument, there are still many elements that are incomplete and which need to be completed because of the specificity of environmental issues.

In the decision-making process, integrated physical planning corresponds to the development phase which fits between the definition of options listed in a development plan and the programming of pinpoint projects. It is clear that ecological considerations are indispensable at this level. The overall analysis of ecosystems, taking into account the dynamics of the environment, seems to constitute the method of study most likely to provide information on modifications that can affect the human and natural environments. It has been shown that such an overall analysis can be fitted into various institutional contexts at the same time as integrated physical planning. In any case, the practical methodologies and modalities of their application do not currently permit efficient intervention throughout the whole process: initial option, drawing up the project, execution and follow-up. Integrated studies of the ecological-planning type can, however, contribute information on elements of the natural system under consideration and the interrelationships of these elements, combining the objectives and means used in a given development project. In combination with more precise environmental impact studies, they should provide the basic information necessary for understanding the environment in its many dimensions and preserving the possibility of a choice between alternatives.

CHAPTER 11

Transboundary Environmental Issues in the US-Mexico Borderlands

Niles Hansen

ENVIRONMENTAL protection is a worldwide concern. Yet ever since the creation of the United Nations Environment Programme, it has been apparent that the relative priorities of the developed and developing countries tend to differ on this subject. The industrial countries have been the major users of the world's natural resources and, in the process, they have become the heaviest dischargers of environmental pollutants. But they have also recognized the harmful consequences of pollution and, being wealthy, are in a position to "do something" about the problems. In contrast, if they are to improve the material well-being of their people, many developing countries must achieve higher levels of economic development; the diversion of scarce resources from development to environmental protection appears to them to be an extravagant luxury when so many people live at (or not far above) subsistence level.

The rapidly urbanizing, US-Mexico border area represents a unique interface between an advanced industrial nation and a nation that is advancing economically but still shares many of the values and attitudes of the Third World. A critical examination of environmental issues in this situation is the purpose of this chapter.[1] The nature of urbanization in the Mexico borderlands is discussed, followed by a consideration of the transboundary air and water pollution problems. Finally, it is argued that although the economic theory of transboundary pollution issues is well developed, its relevance for policy implementation requires shared values and standards, which are often difficult to find when North and South meet.

[1]This research was supported by the Andrew W. Mellon Foundation and the Mexico-United States Border Research Program, University of Texas at Austin.

GROWTH OF THE US-MEXICO BORDERLANDS

The US Side

The present-day boundary between Mexico and the United States resulted from the military conquest of Mexico by the United States in the War of 1846-1848. The imposition of this demarcation line in 1848—it has been modified only slightly since then—meant that Mexico lost about half of its territory to an alien people who were aggressively pushing westward under the banner of Manifest Destiny. However, at that time few people lived in what was to become the US Southwest; and of the 80,000 Mexicans who did become instant Mexican Americans, fully three-fourths were concentrated in the upper basin of the Rio Grande River, well away from the border. The great westward surge of population following the Civil War tended to bypass the borderlands, which, in terms of non-Hispanic settlement, has largely grown from west to east over time.[2] Indeed, until the coming of the railroads in the 1880s, both sides of the border were very sparsely populated.

Table 1: Population Change (1970-1980), Hispanic-Origin Population (1980) and Birth Rates (1975) in the United States Southwest Borderlands/Standard Metropolitan Statistical Areas (SMSA)

Metropolitan Area	*Population (Thousands)* 1970	1980	Population Change, 1970-1980 (percentage)	Hispanic Population, 1980 (percentage)	Birth rate, 1975[a] (percentage)
San Diego CA	1,357	1,861	37.1	14.8	14.9
Tucson AR	352	531	51.1	21.0	16.4
Las Cruces NM	70	96	38.1	52.1	19.3
El Paso TX	359	480	33.6	61.9	21.6
Laredo TX	73	99	36.2	91.5	27.2
McAllen TX	182	283	56.0	81.3	28.6
Brownsville TX	140	210	49.4	77.1	27.1

[a]Births per thousand of population.

Sources: For columns 1-4, US Bureau of Census (1980) and for columns 5, US Bureau of the census (1978).

[2]This phenomenon is still reflected in the ethnic composition of the US border metropolitan areas; see the data on Hispanic population in Table 1, where the respective areas are listed in west-to-east order.

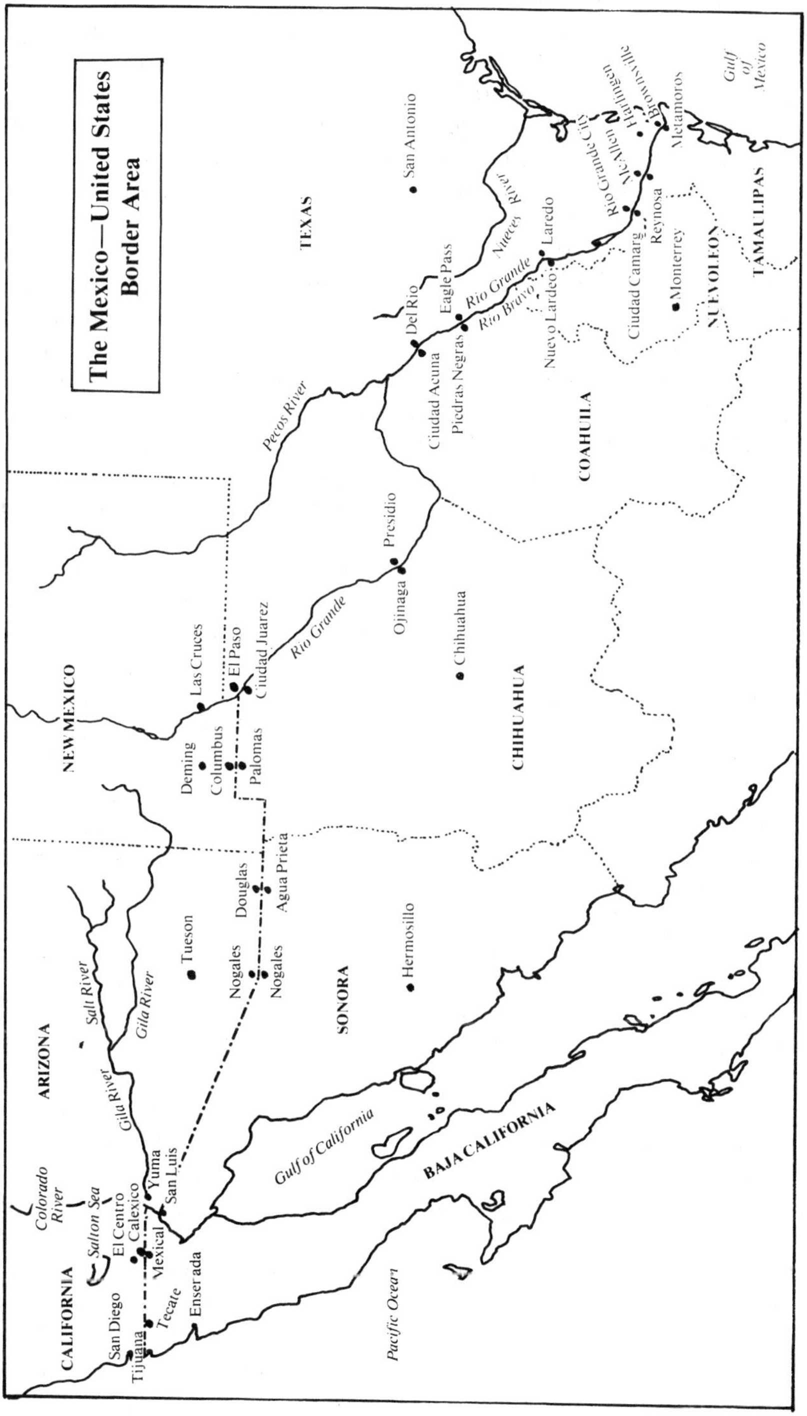

Figure 1: The Mexico - United States Border Area

Today, the seven Standard Metropolitan Statistical Areas (SMSAs), shown in table 1, account for 90 per cent of the total US border population. The San Diego SMSA, which accounts for nearly half of this population, grew relatively rapidly once it became accessible. Even so, as recently as 1910, San Diego was still essentially an agricultural and resort community with fewer than 40,000 inhabitants. Its favourable climate, large naval base and aircraft-construction industry have assured its expansion since then. Tucson, the next largest border SMSA, began as a major mining, agricultural and commercial centre for southern Arizona. It is the home of the University of Arizona and has become increasingly important as a manufacturing and resort area.

In each of the five other US-borderland SMSAs, Hispanics account for over half of the local population, ranging from 52 per cent in Las Cruces to over 91 per cent in Laredo. The largest of these SMSAs is El Paso, a major military, educational and tourism centre. El Paso also has more than 400 industrial plants, including oil refineries, a copper smelter and a copper refinery, cement plants, cotton and food processing facilities and numerous clothing manufacturers. Government-related employment has been a major factor in the growth of the Las Cruces SMSA, which contains New Mexico State University and a portion of the White Sands Missile Range. The Laredo SMSA is strategically located on the most direct route from the United States to Mexico. Laredo is the largest inland port of entry into the United States and moreover, its per capita retail sales are the highest in the nation, largely because of purchases made by Mexican citizens. The McAllen and Brownsville SMSAs are both located in the Lower Rio Grande Valley, where agriculture is the mainstay of the economy. In addition, Brownsville has the most important shrimping industry in the United States. Food processing was the first major manufacturing activity in the Valley and still remains the most important. Apparel manufacturing is the second largest industry. The Valley provides a home base for thousands of Mexican-American migrant farm workers and South Padre Island, on the Gulf of Mexico, has been experiencing a boom in the construction of vacation hotels and second-home condominiums.

During the 1970s, every US-borderland SMSA grew more rapidly than in the previous decade and, with the exception of San Diego, the increase in each case was dramatic. El Paso, which had the lowest 1970-1980 borderland SMSA growth rate, still grew at over three times the corresponding national rate of 10.8 per cent. The McAllen SMSA's growth rate of 56 per cent was five times that of the United States and the Tucson and Brownsville SMSAs grew almost as rapidly. Of the 279 SMSAs that were officially defined as of 1980, McAllen ranked eleventh in terms of growth during the 1970s; Tucson ranked fifteenth and Brownsville, eighteenth. In addition to net immigration, high birth rates have also contributed to population growth in the US borderlands. As shown in table 1, the birth rate in 1975 in every borderland SMSA exceeded the corresponding national rate of 14.8 per thousand of population. Differences in birth rates along the border are directly associated with the proportion of Hispanics in the local population, as can be seen in the data presented in table

1. Thus, while the San Diego and Tucson birth rates did not differ much from that for the nation, those in Laredo, McAllen and Brownsville were about twice the national rate.

The Mexican Side

For decades following the War of 1846-1848, Mexico deliberately ignored its northern states for fear that their development could induce further expansionist moves by the United States. As recently as 1921, Tijuana and Mexicali, both on the border with California, had populations of 1,000 and 7,000 respectively. In 1940, they still had modest populations of 17,000 and 19,000, respectively. But by 1970, these two cities as well as Ciudad Jaurez, on the border with Texas, ranked among the ten largest cities in Mexico. And by 1980, these three cities accounted for half of the 3.4 million persons living in Mexico's northern border area (table 2).

Table 2: Population Change in Northern Border Area Municipios of Mexico (1970-1980).

Municipio	Population 1970	1980[a]	Population change 1970-1980 (percentage)
Tijuana	340,583	639,921	87.9
Mexicali	396,324	516,308	30.3
San Luis R.C.	63,604	99,059	55.7
Nogales	53,494	74,228	38.8
Juarez	424,135	578,471	36.4
Nuevo Laredo	151,253	238,470	57.8
Reynosa	150,786	244,913	62.4
Matamoros	186,146	242,973	30.5
Other municipios[b]	392,337	669,405	70.6
Total	2,158,662	3,404,018	57.7

[a]The 1980 figures are estimates.
[b]Includes all municipios that had a population of fewer than 50,000 persons in 1970.
Source: Mascorro (1981).

Urbanization on Mexico's northern border has been strongly conditioned by the proximity of the United States and it has been both a cause and effect of migration from the interior. Minimum wage rates in Mexico vary by state and by economic subregions within states. The subregions on the border have relatively high rates: Baja California Norte has the highest in all Mexico (Schlagheck, 1977), as well as the highest median monthly earnings (Greenwood, 1978). Even so, wage rates on the Mexican side of the border are considerably lower than those on the US side, which in turn are low in relation to the United States as a whole. Thus, the growth of Mexico's border cities has been partly a result of opportunities to work across the border. These cities have served as a staging ground for persons who work in the United States on an undocumented basis and they have attracted many persons who live in Mexico but cross the border regularly to work in US border cities. Commuters and other Mexican border residents also do much of their shopping on the US side. US residents, in turn, flock to the Mexican cities for tourism, entertainment and bargain-hunting. Indeed, on balance, border transactions have traditionally been favourable to Mexico. All along the border, twin cities exist in a highly symbiotic relationship.

In response to high unemployment rates, the Mexican government has been attempting to attract US plants to the Mexican border area (Hansen, 1981a). The assembly plant, or *maquiladora,* programme takes advantage of US tariff code provisions that allow foreign-based subsidiaries of US firms to assemble products whose parts were originally made in the United States and then to export the products to the United States, with duties being imposed only on the value added (cheap labour). In keeping with the new international division of labour, many US firms have participated in this programme, which currently provides some 120,000 jobs for Mexicans. The programme has been criticized for increasing Mexico's dependence on the United States but Mexico has, nonetheless gained employment, foreign exchange and an expanded industrial base for its border cities.

In order to discourage growth in Mexico City[3] and to promote industrial growth in medium-size metropolitan centres with economic development potential, the Mexican government has formulated a National Urban Development Plan. Among the ten areas chosen for this initiative are three along the US-Mexican border: the Baja California - Sonora system will cover the territory from Tijuana to Nogales; the other two areas are opposite the Texas borderlands: the Ciudad Juárez - Ciudad Acuna system and the Piedras Negras - Nuevo Laredo - Matamoros system (Secretaría de Programación y Presupuesto, 1978; Southwest Border Regional Commission, 1980). To the extent that these efforts are successful, added impetus will be given to the attractiveness of Mexico's northern borderlands, where the overall rate of population growth already exceeds even the remarkably high growth rates for

[3]The Mexico City metropolitan area accounts for 25 per cent of Mexico's total population, 42 per cent of the employed workers, 38 per cent of the value-added in industry, 75 per cent of the professional schools, 58 per cent of the automobiles and 49 per cent of household durable goods sales (Olmedo, 1981).

Mexico as a whole and the US borderlands.

A discussion follows on the environmental consequences of this rapid growth in the US-Mexico borderlands, together with the responses that have been attempted within this unique international setting.

TRANSBOUNDARY ENVIRONMENTAL ISSUES

Air Pollution

Environmental quality in the borderlands is adversely affected by population growth, industrial development and agricultural activities. Comprehensive data concerning air pollution in the region are not available but there is considerable partial evidence concerning problems that necessarily have an impact on both sides of the border.

Industrial activities contribute to air pollution through fuel combustion or manufacturing process losses or residues. The US Clean Air Act of 1970 required standards to be set and emission inventories to be made for six major air pollutants: particulates, sulphur oxides, nitrogen oxides, total hydrocarbons, photochemical oxidants (reactive hydrocarbons) and carbon monoxide. Point-source estimates of air-pollution emissions in US border counties were made in 1970 (Schoults, 1974). In terms of tonnage, sulphur oxides (1.01 million metric tons per year) were by far the most prevalent form of pollution. Smelters located from El Paso to Ajo, Arizona, were primarily responsible for these emissions, to which must be added those from Cananea, which produces most of Mexico's copper. Smoke from the Cananea reduction works, located just 35 miles south of the Arizona border, can often be seen drifting to the US side of the border (Enriquez, 1973). The sulpher oxide emissions on the US side alone represented about 19 per cent of the estimated national total.

Particulate emissions (171,000 metric tons) ranked second among pollutants in the US borderlands: rock, sand and gravel quarrying, cement-making and asphaltic concrete batching were the major contributors. Carbon monoxide (123,000 metric tons) ranked third, with the principal contributors being petroleum refining, and chemical and petrochemical processing. Nitrogen oxides (68,000 metric tons) came mainly from fuel combustion in industrial and power-generating operations. Hydrocarbons (32,000 metric tons) were produced largely by chemical processing industries and petroleum refineries. Other air pollutants not measured in the 1970 study but of major significance, (particularly in the El Paso - Juárez area) are heavy metals such as lead and cadmium (Busch, 1978).

Agricultural activity also results in disturbances of the borderlands' fragile ecosystem. For example, intensive farming in the Imperial, Mexicali and Lower Rio Grande Valleys, and dirt roads all along the border, are major causes of high levels of suspended particulates. Brownsville, San Benito,

McAllen and Progresso in the Lower Rio Grande Valley have all been classed as non-attainment areas with respect to this type of pollution. The widespread use of pesticides and herbicides, often applied by aerial spraying or dusting, produces residues that are dangerous to livestock and humans. Persons in the Lower Rio Grande Valley have, in fact, been found to have unacceptable levels of pesticides in their bodies. Other significant sources of pollution from agricultural operations include burning of agricultural wastes, the use of burners in orchards for heating and wind erosion. Irrigation ditches are burned to clear them of vegetation and cotton-gin wastes are often burned both to eliminate trash and to control pest larvae (Busch, 1978; Utton, 1981a, p. 16).

Motor vehicles are another major source of air pollutants. In the mid-1970s, over 1,400,000 vehicles were registered in US border counties and over 200,000 in the border zone of Mexico (Busch, 1978); both figures are no doubt considerably larger today. Although the Mexican side accounts for a relatively small share of all borderland vehicles, the average age of Mexican vehicles is high and frequently they have no air-pollution-control devices on their engines. Carbon monoxide, hydrocarbons and nitrogen oxide emissions pose particular problems in the densely settled, urban areas along the border. The El Paso - Juárez area is a case in point. There are about 300,000 vehicles in El Paso; the number in Juárez has been variously estimated between 28,000 and 160,000, with the true figure probably lying towards the upper end of this range. El Paso is officially classified as a non-attainment area with respect to carbon monoxide and ozone emissions — a new classification used for hydrocarbons usually associated with vehicular emissions (Bath and Rodriguez, 1981, pp. 6-7). The border itself contributes directly to this situation because of the stop-and-go traffic at the ports of entry. Bath and Rodriguez (1981, p. 7) report: "Since it is estimated that approximately twenty minutes is required to pass through customs on the US side of the border and most people leave their cars running, large amounts of carbon monoxide can be found along the border. Ironically, one federal agency, the Customs Service, leads to concentrations of carbon monoxide which another federal agency, the Environmental Protection Agency, says violates federal standards. El Paso is certainly paying the price for being a border city."

In general, and in contrast to the rest of the United States, the air quality in El Paso has been deteriorating; though corresponding data for Juárez are difficult to obtain, the situation there is no better and may even be worse. Meanwhile, the international boundary provides a convenient excuse for each side to blame the other for major responsibility for pollution (Bath and Rodriguez, 1981, pp. 9-10; Barojas-Weber, 1978). Similar conditions exist elsewhere along the border, especially in the San Diego - Tijuana basin, where it is common for the prevailing winds to carry pollutants from Tijuana to San Diego during the morning hours, whereas during the afternoon the opposite phenomenon occurs. Moreover, as in the case of El Paso - Juárez, it is not uncommon for temperature inversions to prevent the dispersal of pollutants

(Martínez and Deschamps, 1981, pp. 7-8).

Finally, pollution in the workplace has become an issue in the Mexican borderlands. It has been argued that some of the *maquiladoras* (assembly plants) have located on the Mexican side of the border in order to escape controls and regulations that protect US workers. Thus, among the electrical, chemical and automotive *maquiladoras* in Tijuana, ten allegedly have been polluting in terms of lead emissions, four in terms of cadmium, two in terms of asbestos and one in terms of mercury (Martinez and Deschamps, 1981, p. 10). Similar allegations have been made concerning plants in Ciudad Juárez (American Friends Service Committee, 1979-1981).

Surface Water Quality

Water scarcity and water quality are increasing problems throughout the south-western United States and northern Mexico. The Colorado River has been the focus of this controversy, both within the United States and between the United States and Mexico.[4] In addition to the demands made by Mexico, Blundell (1979) states that the Colorado River "helps slake the thirst of 11 million people from Denver to San Diego. Diverted into canals and ditches, it waters 3.4 million acres producing crops worth well over US$1 billion yearly. Forced through turbines, it sends electricity leaping out of canyon power plants to more then 200 utilities, rural electric co-ops and others. The only substantial source of surface water in the Southwest, it is relied upon as few rivers are anywhere."

Unfortunately, the quality of Colorado River water has been deteriorating and total legal claims exceed the average flow. Under the terms of the 1922 Colorado River Interstate Compact, the Upper Basin states of Utah, Wyoming, New Mexico and Colorado agreed that half of the then-average water flow — 15 million acre-feet — would be guaranteed to the Lower Basin states of Arizona, California, and Nevada. In 1944, Mexico, by treaty with the United States, received an annual entitlement of 1.5 milion acre-feet, bringing the total amount of legal claims to 16.5 million acre-feet. But the division of Colorado River water was based on a brief and unusually high rate of flow; since 1922 the actual flow has averaged only 13.5 million acre-feet. Moreover, Californa has been 'borrowing' more than its share, which has prompted the Upper Basin states to push for the completion of water resource projects in their area so that the Lower Basin will not continue to overdraw its allotment. A further complication involves the water rights of Indian reservations, which were completely ignored when the 1922 Compact was drawn up. Recent Supreme Court decisions have guaranteed water supplied to reservations for

[4]The Colorado River and its upper tributaries rise in the mountains of Wyoming, Colorado and Utah, where precipitation, primarily in the form of snow, is heavy. The river flows about 1,440 miles to the Gulf of California. Its drainage basin covers 244,000 square miles and includes parts of seven US states. For 17 miles if forms the border between Arizona and Mexico; it then flows 80 miles through Mexico to the Gulf of California.

irrigation and the Navajos, in particular, could conceivably absorb a substantial share of all Colorado River water. So far, the Navajos have not pressed a lawsuit but they have made a survey of irrigable land, the result of which have not yet been released.

One of the biggest drains on Colorado River water will be the Central Arizona Project, which will deliver 1.2 million acre-feet to Tucson and Phoenix by the mid-1980s. Congress has authorized an initial outlay of US$1.6 billion for this massive undertaking and tunnels are already being bored through the Buckskin Mountain to construct the first link of a 300-mile aqueduct. As has been the case with other Colorado River projects, the users of the water will no doubt pay only a fraction of the true costs of supplying it, which in turn provides little incentive for conservation.[5]

In 1979, controversy raged about whether Mexico should pay for damage to Texas beaches, resulting from a massive oil spill from a Mexican-leased rig in the Gulf of Mexico. This issue became linked to a dispute between the United States and Mexico over Colorado River water. Mexican President López Portillo said that Mexico would not pay for oil-spill damage because the United States had not paid for damage to the Mexicali Valley,[6] caused by salty water coming from the Arizona side of the border. The history of this issue goes back to 1961. Before then, Mexico received more Colorado River water than the 1.5 million acre-feet guaranteed in the 1944 treaty. This excess flow served to dilute saline, irrigation return flows from the United States to the point where the quality of the water going to Mexico was about that of the water used in Arizona. In 1961, however, the Wellton-Mohawk Project of the Bureau of Reclamation began operating a system of drainage wells that discharged saline water into the Colorado River below the last US diversion point. At the same time, the United States began storing water in Lake Powell, the reservoir for the almost completed Glen Canyon Dam, and the average flow to Mexico dropped to 1.5 million acre-feet. In consequence, for a period in 1962, the salinity of water delivered to Mexico was one and a half times higher than it had been in 1960. Mexican farmers reacted vigorously to the adverse effects on crop yields and the Mexican government lodged formal protest to the United States (Oyarzabal-Tamargo and Young, 1978; Holburt, 1978).

During a 1972 visit to Washington, Mexican President Echeverría termed the Colorado River salinity problem "the most delicate bilateral problem" between the two countries (Wiessler, 1979). This set the stage for a 1973 agreement by which the United States set out steps to reduce the Colorado's salinity and agreed to "provide non-reimbursable assistance for those aspects of the Mexicali rehabilitation programme that related to salinity" (*ibid.*). No

[5]For example, farmers in the Imperial Valley of California pay $4.50 per 326,000 gallons of water that has been stored behind federal dams and piped to them at enormous cost, whereas San Francisco residents pay the same amount for only 2,460 gallons of unsubsidized water (Boeth *et al.*, 1978).

[6]In 1972, Colorado River water irrigated 203,000 hectares in the Mexicali Valley. This represented over 30 per cent of all irrigated land along the Mexican border area. The Lower Rio Bravo (Rio Grande) irrigation district accounted for about 40 per cent of the total (Revel-Mouroz, 1979).

amount of damages was specified in the agreement and Mexico has never officially estimated the costs, although it could still produce a claim. Meanwhile, interim measures have been taken to solve the salinity problem. Residual irrigation drainage water is now moved west in a channel and discharged directly into the Gulf of California; water removed from the Colorado River for irrigation is currently replaced with clear water from the Hoover Dam. By 1983 or 1984, a desalinization plant will be in operation so that water used in irrigation can be returned in pure form to the river. The cost of these efforts is estimated at US$350 million, and mounting because of inflation. None of this benefits US users directly, although additional salt-control projects are underway upstream for them (Blundell, 1979).

Although Mexico's allocated share of Colorado River water is a prior obligation under the terms of the 1944 treaty, the United States retained the right to reduce deliveries to Mexico in the same proportion as US consumption is reduced in the event of 'extraordinary drought or serious accident' in the country. While the imprecision of this clause still arouses some concern in Mexico, the more important issue from the perspectives of both countries remains that of water salinity (Utton, 1980, p.2). In particular, the Colorado basin is one of the richest storehouses of energy resources in the United States, its water being used in all aspects of energy production including mining, reclamation of mined lands, processing, transportation, refining and energy conversion. Critical issues are raised by the fact that anticipated uses of Colorado River water already exceed the supply and by the fact that salinity concentrations are more affected by taking water that serves for dilution than by the removal of salt from the water. Clearly, as Bishop (1977) states, these "effects of future development will need to be considered in the planning and implementation of programmes to meet water quality and quality commitments to Mexico."

The Rio Grande River, which over a course of 1,255 miles forms the boundary between Texas and Mexico, has been relatively free of pollution problems. At one time, increasing salinity was a threat in the area of the Lower Rio Grande Valley where major diversions are made to irrigated lands in both the United States and Mexico. Between 1966 and 1969, a project was undertaken entirely on Mexican territory to divert saline waters to the Gulf of Mexico. Mexico was responsible for the construction of the conveyance channel but the actual cost was shared by both countries because the salinity was of natural origin and the project benefited users on both sides of the border. Costs of operating and maintaining the diversion works continue to be equally shared (Friedkin, 1981a, p.11). The potential detrimental effects of rapidly increasing population and economic development along the Rio Grande are of concern to both the United States and Mexico and Hernandez (1978) states: "It is important that analytical studies on historical water quality records be carried on so as to detect changes in quality before serious adverse effects occur."

The Santa Cruz River recharges groundwater supplies for the twin cities of Nogales, Arizona, and Nogales, Sonora. The problem of water quality has been met by building an international treatment plant that receives waste water from each city. In 1975, three million gallons per day were treated before discharge into the Santa Cruz (Bradley and DeCook, 1978). A similar facility has also been built for Douglas, Arizona and Agua Prieta, Sonora (Friedkin, 1981b).

In other border areas, substantial surface-flow-quality problems remain. For example, the San Pedro River, which runs northwards from Sonora into Arizona, is frequently discoloured. Farmers in Arizona are concerned about possible damage to their land and crops; residents are disturbed about potential health hazards. The probable source of the contamination is the large copper works at Cananea (Utton, 1981b).

The New River, which rises south of Mexicali and flows northwards to the Salton Sea in California, is perhaps the most polluted stream in the United States. The flow, normally some 3.5 million gallons an hour, consists largely of irrigation drainage and drainage from Mexicali's main municipal dump, untreated slaughterhouse and industrial wastes, and the inadequately treated sewage of over 500,000 people. Local residents along the 55-mile course of the river have long known to keep away from the water. But, because of the increasing number of tourists throughout the Imperial Valley, signs have been erected warning of the dangerous situation. According to California officials, Mexican authorities have been promising for years to install proper sewage treatment facilities in Mexicali and to control other pollution sources, but, says Hill (1978), "they never seem to get around to it."

The Tijuana River, which is about 85 miles long, rises in the highlands of Baja California Norte and flows through Tijuana, before crossing into San Diego County, California, from which point it flows for about six miles to its mouth at the Pacific Ocean. The salt marshes and tidal sloughs of the estuary are totally on the US side, as is most of the flood plain. The estuary is regarded as one of the finest remaining wildfowl sanctuaries in California. During the 1960s, Tijuana officials decided to construct a deep concrete channel to contain the river's normal flow, as well as any likely flood waters; the principal motive was to use the dry river bed for the modernization of the heart of the city. At about the same time, San Diego agreed to build a similar channel, which would have been an extension of the Tijuana Channel from the border to the Pacific. The reclaimed flood plain on the US side was to be used for a variety of developments including hotels, a golf course and a marina. Mexico began the construction of its channel in 1972 and work was completed in 1976. Meanwhile, environmental interests in San Diego launched a campaign to preserve the natural qualities of the estuary. They finally persuaded the San Diego City Council to abandon the channel option (and concomitant economic development plans) in favour of a 'minimum plan' involving a series of earthen dissipator barriers to the flood charge coming from the Tijuana channel. Construction of this system was completed in 1979 but, meanwhile, extensive flooding was experienced in Tijuana. A severe flood in 1980 overtaxed the dissipator system and caused extensive damage on both sides of the border. The net result of the Tijuana River flood-control controversy was a clear victory for the environmentalists, but at the expense of economic development in relatively poor neighbourhoods on the US side and resentment in Tijuana over San Diego's unilateral decision to abandon its original agreement. Current arguments and disagreements with respect to transboundary sewage control continue to reflect the lack of co-operative spirit that has typified relations between San Diego and Tijuana (Duemling, 1981; Kennedy, 1978).

Groundwater Quality

The legal and economic context along the border is such that each nation is encouraged to outdo its neighbour by developing its groundwater resources as rapidly as possible, perhaps even to the point of depletion (Utton, 1980, pp. 9-13). Groundwater in underground basins, located partly in the United States and partly in Mexico, has not been apportioned between the two countries.[7] The Mexican government, through the Secretariat of Water Resources, has the legal authority to regulate water extraction and establish prohibited ground water zones, if existing developments or the aquifer are in danger of being adversely affected or if it is otherwise in the public interest (Utton, 1980, p.9). The four US border states are the heaviest users of groundwater in the nation and shortages in the border region intensify the water supply problem. Each of these states has a different system of groundwater law; none has adequate legislation or regulations for the protection and management of diminishing supplies within the state and along the border areas. New Mexico has the only public control system but its regulations do not contemplate joint controls in the international boundary area. Arizona and Texas have no controls other than those of a voluntary nature and Californian law does little to discourage excessive pumping and waste (Clark, 1978).

The environmental effects of the non-management of groundwater resources are perhaps best illustrated by the situation in the El Paso - Juárez area. Both cities rely almost exclusively on the Hueco Bolsón for their water needs. The Hueco Bolsón is an artesian aquifer and there is a freshwater - saline water interface that has been maintained by aquifer pressures. El Paso and Juárez are each pumping water at a rate more rapid than that at which the groundwater reservoir is being recharged.[8] Between 1903 and 1976, water levels fell as much as 73 feet in the centre of El Paso and 85 feet in Juárez. By 1991, the annual recharge may be as little as 5 per cent of the annual

[7]However, the two governments have approved a mutual pumping limitation in a particular problem area along the border near the Colorado River, pending a comprehensive agreement on groundwaters along the boundary (Friedkin, 1981a, p.9).

[8]*Per capita* groundwater use in El Paso has been about three times that in Juárez (Day 1978, p.175). Water storage and distribution facilities in Juárez are not as developed as those in El Paso. Informal settlements have particular difficulty obtaining potable water. On one hot day in 1981, at least twelve persons in Juárez died from lack of water and another 1,500 suffered from dehydration (Aguilar, 1981).

withdrawal. The amounts of freshwater that can be removed from the Hueco Bolsón under 'safe yield' conditions have not yet been determined. But there is evidence that depletion is already causing some groundwater degradation due to the encroachment of saline water from adjacent sands (Utton, 1981a, pp. 9-10). Through the co-ordinating efforts of the International Boundary and Water Commission, the United States and Mexico have exchanged data concerning water appropriations from the Hueco Bolsón and the national and international implications of the demands on this resource. However, Texas authorities have been unenthusiastic and sometimes even indignant towards this initiative (Day, 1978, pp. 176-177). Texas law implicitly acknowledges that underground water will eventually run out (if it is not first rendered unfit for use by salinity) by granting farmers a depletion allowance. But Texas law has also traditional landowners, and these seem more interested in racing each other to the bottom of the aquifer than in even considering the possible long-term benefits of co-operative management strategies.

INTERNATIONAL ECONOMIC INEQUALITY AND ENVIRONMENTAL PROTECTION

Economic Theory

In theoretical terms, the economics of transboundary pollution has been elaborated in considerable detail (OECD, 1974 and 1976). Only a summary of the general principles will be attempted here. If common emission standards can be agreed upon, then each nation is merely required to enforce compliance in accordance with the 'polluter pays principle'. If agreement takes the form of common environmental quality standards, then each nation must formulate an appropriate combination of standards for various harmful emissions so that the objectives are achieved, again in keeping with the 'polluter pays principle.' Without an agreement on standards, this principle could still be used to determine the allocation of control costs if the countries could agree on an optimum point to which pollution should be abated. Joint pollution should be abated to the point where combined marginal damage costs equal combined marginal control costs. Abatement would be allocated between the countries, according to the marginal cost curves of each one and the combined marginal damage cost curve. The international optimum would be achieved where these curves are equated in each country.

While these options are relatively reasonable when the countries concerned are at similar levels of development and share common values, they are not likely to be relevant where international inequality is pronounced. For example, the adoption of common standards would imply that the relatively poor country would have to devote a higher proportion of its resources to pollution abatement than would the relatively rich country. Apart from abstract questions of justice, this is not a circumstance that would lend itself to agreement between the two countries. Moreover, the approach that would seek a mutually determined optimum level of pollution would require that each country concur that international resources are, in some sense, internationally owned and also agree to accept the validity of the other country's marginal control cost curves and damage cost curves. In addition, it must be agreed that the cost functions can be combined or else the optimum is indeterminate. But, if opportunity costs are taken into account, then according to Marquand (1976): "The marginal control costs for a given abatement of pollution would often be much higher for a poor country or indeed for any country with an unemployment problem, than for a country where few resources were idle and where public preferences were relatively stronger for a cleaner environment and less strong for more goods. The poor country would tend to want to abate less than the rich country. Similarly, different countries are likely to value differently the damage which pollution causes them."

The marketing of pollution rights is probably a relatively efficient means for dealing with intranational environmental problems. It has also been proposed for the US-Mexican borderlands (Ayer and Hoyt, 1977). However, this approach, particularly where international economic inequality is great, also runs into the problem of deciding how to proceed where different countries have different preferences and where no agreement on commission standards is reached (Marquand, 1976).

Conflicting Values

The difficulties just outlined are quite apparent in the US-Mexican borderlands. On paper, Mexico has one of the world's most impressive legal regimes for environmental protection but, as in many Third World countries, the Government *de facto* accepts pollution as part of the price that must be paid for economic development. Juergensmeyer and Blizzard (1973) state: "The reconciliation of a need to develop new industries and protect existing ones from foreign competition and the need to acquire costly equipment and processes in order to protect the environment is difficult indeed. Furthermore, the expenditure of public monies for environmental projects designed to enhance beauty or convenience is difficult to justify in a society with so many citizens at the poverty end of the economic spectrum. Every indication is that, in spite of the forcefulness of the new laws, environmental protection in Mexico will be subordinated to frequently inconsistent policies of economic development."

Although 'ecoplans' have been prepared for all of Mexico's border states, they are in fact general statements of desired goals, lacking specific details concerning implementation. Local capability for dealing effectively with environmental problems in Mexico's border cities is virtually non-existent; levels of air and water pollution continue to rise and *maquiladoras* that produce carcinogens continue to be welcomed in border industrial parks. (Bath and Rodriguez, 1981, pp. 16-18).

The situation on the US side of the border is more heterogenous but, here too conflicts between development and environmental objectives are apparent. Most indicators of economic welfare decrease from west to east along the US borderlands (Stoddard, 1978; Hansen, 1981b). In the San Diego SMSA, for example, per capita income is above the national average, whereas the Laredo, McAllen and Brownsville SMSAs are the poorest in the nation. Enforcement of federal environmental legislation has largely been left to the respective states and California has generally been considered relatively strict in this regard. In keeping with this phenomenon, it has been indicated that among US border areas, the constituency on behalf of environmental causes has been relatively strong in San Diego. In contrast, the situation in the Texas borderlands has been quite different. Here, environmental pollution "as a problem ranks far behind others such as unemployment, health care, industrialization, lack of development and a host of others" (Bath, 1978, p. 183). This attitude is reflected in the lax regulatory efforts of state and local officials (Bath and Rodriguez, 1981, pp. 10-15). It may thus be said that in moving from west to east, the degree of environmental concern and the degree of actual environmental management both tend to decline. In San Diego, environmentalists have so far triumphed over economic development proponents when their interests have clashed; in El Paso, there is sporadic concern about the environment but it is not sufficiently sustained to bring about effective management of environmental resources; in the Lower Rio there is little evidence even of concern.

Local officials in twin cities along the Texas-Mexico border generally recognize their interdependence and have developed extensive, but informal, networks for dealing with such transboundary issues as law enforcement, social festivities, information and equipment exchange, public health, fire control, banking and tourism (Sloan and West, 1976 and 1977). Thus, their lack of co-operation in environmental matters is largely a reflection of relative disinterest in the topic. In contrast, whereas San Diego clearly has a strong environmental movement, it has a poor record with respect to its dealings with Tijuana. San Diego's growth has been fueled by an influx of Mid-westerners and Southerners, and as Duemling (1981, p. 26) reports: "Most would have happily 'pulled up the drawbridge' once they were inside. With no interest in what transpired across the border, these people reinforced San Diego's historical parochialism. The community was thus very slow to recognize the implications of explosive growth in Tijuana, including threats to public health and the quality of water and air." Although survey evidence indicates that the citizens of Tijuana are better informed about the United States than San Diego residents are about Mexico, it remains true that Tijuana has preferred to go its own way, avoiding any action that would suggest dependency on the United States (*ibid.*, pp.2 and 17).

CONCLUSIONS

Continuing rapid growth on both sides of the US-Mexico border will no doubt intensify the many environmental problems already present in this area. In the past, economic development issues overshadowed environmental concerns on the Mexico side as well as in the Texas borderlands, which is a relatively poor region by US standards. In the more prosperous western US borderlands, environmental quality has received much more attention but some unilateral actions in this regard have led to conflicts with the more development-oriented Mexico side. Moreover, all along the border, the political boundary has represented an excuse for each side to blame the other for environmental pollution. Such progress as has been achieved in limiting pollution (for example, improved pesticide management and the reduction of sulphur oxides from copper smelters) has generally not involved international co-operation.

The need for better and more comprehensive management of environmental resources along the border has been repeatedly stated in much of the relevant literature. It has frequently been suggested that the International Boundary and Water Commission be given expanded powers, particularly with respect to air-pollution control. However, the Commission has been oriented towards a rather restricted range of purely technical tasks; it has had no contact with pollution problems and "apparently does not want any contact with air-pollution regulation" (Bath, 1978, p. 190). This does not seem to be a promising basis for expanding transboundary environmental co-operation. On the other hand, a legal framework for international co-operation was established in 1979, when the Presidents of both countries signed a Memorandum of Agreement prepared by the US Environmental Protection Agency and Mexico's Subsecretaría de Mejoramiento del Ambiente. It contained the following major provisions:

1. A co-operative effort should be made between the two agencies to resolve environmental problems of mutual concern in border areas, exchange information and personnel and establish parallel projects between the two parties.
2. Parallel activities are to be undertaken but joint actions to resolve specific problems may be conducted.
3. SMA and EPA senior officials are to meet annually at alternate sites to discuss overall policies, programmes and problems.
4. Experts designated by the EPA and SMA may meet to review technical issues and plan parallel projects, including pollution abatement and control, quality assurance, research and monitoring. These experts are to meet annually and may make policy recommendations to their respective governments.
5. Officials may expand their discussions to include all areas of environmental concern.

6. Each party will name one person as co-ordinator to facilitate exchange of information and these co-ordinators may invite representatives of local, state, international organizations or private citizens to participate in meetings, conferences and parallel activities.
7. Parallel activities may include pollution abatement and control programmes directed towards specific pollution problems, development of an early-warning system to alert the two governments to potential environmental problems and development of data gathering (Bath and Rodriguez, 1981, pp. 21-22).

It should be noted that 'parallel activities' seem to be preferred over joint efforts, which is in keeping with the historical preferences of both sides of the border to undertake co-operative efforts on a separate basis (Duemling, 1981, pp. 26-27). In any case, concrete results have yet to issue from this agreement, even though it does represent high-level recognition of shared problems and the need for corrective action. The evidence suggests that agreements in principle between national agencies are not likely to yield many tangible results, so long as border communities do not demand action and continue to distrust their respective national governments, as well as, in some instances, their own neighbours across the border.

As the borderland area experiences the qualitative changes associated with economic development—as contrasted with mere growth in the magnitude of historically given variables—there will no doubt be value changes favouring enhanced environmental protection measures. But too often the lack of an ounce of prevention now will mean that a pound of cure will be required in the future. If the specific outcomes and their timings are not yet clear, the evolution of the US-Mexico borderlands should nevertheless continue to be instructive in the world's larger international context.

References

Aguilar, S. (1981). *Doce Muertos y 1,500 Deshidratados por el Calor en Cd. Juárez.* Excelsior, 13 July 1981, p. 26A.

American Friends Service Committee (1979-1981). *Mexico-US Border Program.* Boletin Informativo, No. 9 (November-January 1979-1980), p. 14; No. 10 (January 1981), p. 12; No. 11 (February 1981), p. 5.

Ayer, H. W. and Hoyt, P. G. (1977). *Industrial Growth in the US Border Communities and Associated Water and Air Problems: An Economic Perspective.* Natural Resources Journal, Vol. 17, No. 4, p. 612.

Barojas-Weber, L. H. (1978). *Impacto del Crecimiento en la Calidad del Aire en las Communidades Fronterizas.* Natural Resources Journal, Vol. 18, No. 1, pp. 101-109.

Bath, C. R. (1978). *Alternative Co-operative Arrangements for Managing Transboundary Air Resources along the Border.* Natural Resources Journal, Vol. 18, No. 1, pp. 183 and 190.

Bath, C. R. and Rodriguez, V. (1981). *Comparative and Binational Air Pollution Policy in El Paso, Texas and Ciudad Juárez, Chihuahua.* Department of Political Science, University of Texas at El Paso, pp. 6-7, 9-10, 10-15, 16-18 and 21-22. (Mimeograph).

Bishop, A. B. (1977). *Impact of Energy Development on Colorado River Water Quality.* Natural Resources Journal, Vol. 17, No. 4, p. 669.

Blundell, W. E. (1979). *Colorado River, Vital to Southwest, Travels Ever-rockier Course.* Wall Street Journal, 12 February 1979, p. 1.

Boeth *et al.* (1978). *Western Water Fight.* Newsweek, 12 June 1978, p. 51.

Bradley, M. D. and DeCook, K. J. (1978). *Ground Water Occurrence and Utilization in the Arizona-Sonora Border Region.* Natural Resources Journal, Vol. 18, No. 1, p. 36.

Busch, A. W. (1978). Environmental Management: A Basis for Equitable Resource Allocation. In *Views Across the Border: The United States and Mexico.* Stanley R. Ross (Ed.), University of New Mexico Press, Albuquerque, pp. 352-354.

Clark, R. E. (1978). *Institutional Alternatives for Managing Groundwater Resources: Notes for a Proposal.* Natural Resources Journal, Vol. 18, No. 1, pp. 155-156.

Day, J. C. (1978). *International Aquifer Management: The Hueco Bolsón on the Rio Grande River.* Natural Resources Journal, Vol. 18, No. 1, pp. 175-177.

Duemling, R. W. (1981). *San Diego and Tijuana: Conflict and Co-operation between Two Border Communities.* Executive Seminar in National and International Affairs, Foreign Service Institute, US Department of State, Washington, DC., pp. 2, 17, 26-27.

Enriquez, W. (1973). International Legal Implications of Industrial Development along the Mexican -US Border. In *Pollution and International Boundaries: United States-Mexican Environmental Problems.* Albert E. Utton (Ed.), University of New Mexico Press, Albuquerque, p. 94.

Friedkin, J. F. (1981a). *The International Boundary and Water Commission: United States and Mexico.* Paper prepared for the United Nations Meeting of International River Organizations, 4-14 May 1981, Dakar, Senegal, p. 9 and 11. (Mimeograph).

Friedkin, J. F. (1981b). *United States and Mexico Co-operation on Boundary and Water Matters.* Paper prepared for the UCLA Chicano Research Centre Symposium, 24 April 1981, Santa Monica, California, p. 12. (Mimeograph).

Greenwood, M. J. (1978). *An Economic Model of International Migration and Regional Economic Growth in Mexico.* Journal of Regional Science, Vol. 18, No. 1, p. 18.

Hansen, N. (1981a). *Mexico's Border Industry and the International Division of Labour.* Annals of Regional Science, Vol. 15, No. 2, pp. 1-12.

Hansen, N. (1981b). *The Border Economy.* University of Texas Press, Austin, pp. 142-146.

Hernandez, J. W. (1978). *Interrelationship of Ground and Surface Water Quality in the El Paso-Juárez and Mesilla Valleys.* Natural Resources Journal, Vol. 18, No. 1, p. 8.

Hill, G. (1978). *California Fights River's Flow of Mexican Wastes.* New York Times, 20 November 1978, p. A19.

Holburt, M. B. (1978). International Problems. In *Values and Choices in the Development of the Colorado River Basin,* Dean F. Peterson and A. Berry Crawford (Eds.), University of Arizona Press, Tucson, pp. 220-225.

Juergensmeyer, J. and B., Earle (1973). Legal Aspects of Environmental Control in Mexico: An analysis of Mexico's New Environmental Law. In *Pollution and International Boundaries: United States-Mexican Environmental Problems.* A. E. Utton (Ed.), University of New Mexico Press, Albuquerque, pp. 115-116.

Kennedy, W. C. (1978). *Ecology and the Border: The Case of the Tijuana Flood Control Channel.* Paper prepared for the American Sociological Association Meetings, 4-8 September 1978, San Francisco. (Mimeograph).

Marquand, J. (1976). A note on Some Problems of Transfrontier Pollution. In *Economics of Transfrontier Pollution,* OECD, Paris, pp. 13-14.

Martínez, G. G. and Deschamp, J. R. (1981). *Efectos Ambientales de la Interrelación Fronteriza: El Caso Tijuana-San Diego.* Primer Encuentro Sobre Impactos Regionales de las Relaciones Economicas México-Estados Unidos, 8-11 July 1981, Guanajuato, Mexico, pp. 7-8 and 10. (Mimeograph).

Mascorro, E. V. (1981). *Caracteristicas del Desarrollo Economico de la Franja Fronteriza Norte de México.* Paper prepared for the Coloquio Sobre Crecimiento y Demanda de Recursos en la Zona Fronteriza El Paso-Matamoros, 23 April 1981, South Padre Islands, Texas, p. 6. (Mimeograph).

OECD Organization for Economic Co-operation and Development (1974) *Problems in Transfrontier Pollution.* OECD, Paris.

OECD (1976). *Economics of Transfrontier Pollution.* OECD, Paris.

Olmedo, R. (1981). *Descentralización o Catástrofe.* Excelsior, 13 July 1981, p. 6A.

Oyarzabal-Tamargo, F. and Young, R. A. (1978). *International External Diseconomies: The Colorado River Salinity Problem in Mexico.* Natural Resources Journal, Vol. 18, No. 1, p. 78-79.

Revel-Mouroz, J. (1979). *Frontière politique et frontiére agro-pastorale dans le Nord du Mexique.* Actes du Cent Troisième Congrès National des Sociétés Savantes, Section de Geographie. Bibliothèque nationale, Paris, p. 246.

Schlagheck, J. L. (1977). *The Political, Economic and Labour Climate in Mexico.* Industrial Research Institute, Wharton School, University of Pennsylvania, Philadelphia, pp. 73-76.

Schoults, J. M. (1974). Air Pollution from United States Industrial Sources along the United States-Mexico Border. In *Air Pollution along the United States-Mexico Border.* Howard G. Applegate and C. Richard Bath (Eds.), Texas Western Press, El Paso, pp. 28-33.

Secretaría de Programacíon y Presupuesto, Comisión Nacional de Desarrollo Urbano, Secretaría de Asentamientos Humanos y Obras Públicas (1978) *Plan Nacional dè Desarrollo Urbano.* Mexico City.

Sloan, J. W. and West, J. P. (1976). *Community Integration and Policies Among Elites in Two Border Cities.* Inter-American Studies and World Affairs, Vol. 18, No. 4, pp. 451-474.

Sloan, John W. and West, J. P. (1977). *The role of Informal Policy Making in US-Mexico Border Cities.* Social Science Quarterly Vol. 58, No. 2, pp. 270-282.

Southwest Border Regional Commission (1980). *Economic Development Plan.* Tucson, pp. V-18 and 19.

Stoddard, E. R. (1978). *Patterns of Poverty along the US-Mexico Border.* Center for Inter-American Studies, University of Texas at El Paso.

US Bureau of Census (1978). *County and City Data Book 1977.* US Government Printing Office, Washington, D.C.

US Bureau of Census (1980). *Final Population and Housing Unit Counts.* 1980 Census of Population and Housing, Advance Reports No. PHC80-V-4, Arizona; No. PHC80-V-6, California; No. PHC8-V-33, New Mexico; and No. PHC80-V-45, Texas. US Government Printing Office, Washington, D.C.

Utton, A. E. (1980). *Water Resources Development and the Environment in the Border Regions.* Paper presented at the Mexico-United States University Conference on Border Studies, 28-29 February 1980, La Paz, Mexico, pp.2 and 9-13. (Mimeograph).

Utton, A. E. (1981a). *Summary Report: Working Group on US-Mexican Transboundary Resource Needs and Issues to the Year 2000.* 23-25 April 1981, South Padre Island, Texas, pp. 9-10 and 16. (Mimeograph).

Utton, A. E. (1981b). *Selected Resource Issues in the Border Region.* Paper presented at the First Conference on Regional Impacts of United States-Mexico Economic Relations. 8-11 July 1981, Guanajuato, Mexico, p. 8 (Mimeograph).

Wiessler, J. (1979). *Mexico was offered pay but did not bill US for Salt Water Damages.* Houston Chronicle, 7 October 1979, p. 6, sec. 1.

CHAPTER 12

Conclusions and Recommendations

THE integrated physical planning approach, which subsumes socio-economic and environmental considerations, has been considered by some as a prescription for society's need for long-term and substantial development, by others as a jargon-ridden overstatement of the obvious. In fact, it is neither. It is correct to say that integrated planning methods have clarified thoughts, classified objectives and improved the available planning techniques. However, the enormous complexity of the system which confronts environmental managers and decision-makers, under present conditions, places enormous obstacles to the development of practical planning procedures and modalities. At the same time, without such planning procedures and modalities, our ability to understand, manage, control and forecast environment/development relations must remain incomplete and tentative. At the very least, integrated planning can be viewed as a complex training ground for developing management skills, required in dealing with escalating environmental problems. Despite the success of the application of integrated planning to the socio-economic and environmental systems, particularly at the technical level, it is obvious that difficulties still remain in the way of a cogent application of integrated planning techniques to the solution of a wide range of significant societal problems.

THE NATURE OF INTEGRATED PLANNING

The dynamism of society calls for planning methods to cope with changing social values and the increasing scale of economic and technical development. Integrated planning has been proposed as an organizational arrangement to deal effectively with the alternatives and options of an increasingly complex spectrum of socio-economic and environmental change and development. It has been realized that policies designed to solve one issue affect not only the context within which it is being considered, but also the range of actions on other issues. Consequently, integrated planning has to be viewed as a multiple-objective-oriented activity, based on an understanding of the interrelationships between the various subsystems. Thus, integrated planning is a framework for analysis and identification of interdependent factors at work in any particular socio-economic and environmental situation. Integrated planning is neither a static entity nor a perfect concept of rationality but, merely, a more adequate approach to public policy making.

Definition

Integrated planning can be defined as a set of procedures required to achieve the objectives determined by a particular society. It is not a single act, but rather a complex of procedures undertaken, simultaneously or sequentially, to accomplish a socially acceptable outcome. In this context, integrated planning can be distinguished firstly, as a framework to thought because it requires an overall assessment of the various interdependent issues and problems involved and, secondly, it can be viewed as a technique for policy-making by means of which complex and synergetic problems can be solved.

Objectives

A fundamental methodological problem for decision-makers is the lack of an explicit statement of trade-offs between socio-economic and environmental objectives. Policy options (such as the 'best' combination of, for example, the level of inflation and unemployment in economic policy or economic efficiency versus environmental quality in environmental policy) are not determined by planners since they represent political value judgements. Only in a few cases are the decision-makers aware of the trade-offs involved. Equally, serious problems exist in interdisciplinary integration; that is to say, economic and environmental decision-makers operate with different criteria. Economic planners are predominantly concerned about 'efficiency' and to a lesser degree with 'equity', while environmental planners give more consideration to 'diversity' and 'stability'. Thus, a common denominator among the different disciplines is required. Efforts to integrate, for example, economic and environmental planning will force decision-makers to state explicitly their criteria and their policy options. In essence, the integrated planning approach tries to accomplish the following objectives:

(a) to accomplish interdisciplinary equity in order to establish a balance between conflicting criteria, interests and objectives;

(b) to increase efficiency of the overall physical, socio-economic and environmental system in order to achieve an optional level of quality of life; and

(c) to develop public participation with the purpose of democratizing the decision-making process and to make it more socially acceptable.

Thus, integrated planning refers as much to an attitude of mind concerning societal problems as it does to planning techniques with prescribed methods and procedures.

Institutional Arrangements

The existing governmental institutions and structures in many countries, with decision-making functions distributed throughout sectorial agencies, do not facilitate integrated planning. There is a need for restructuring the bureaucracies for this purpose. In addition, there are no universally applicable analytical methods of integrated planning.

One possible approach to integrated planning, available for countries with sufficient manpower and resources, is a policy of learning through experimentation. However, a key problem with such a policy is how to prepare bureaucracies that one is in the process of learning. Furthermore, no individual agency gets involved in totally integrated planning because none (even national governments) can provide the total available influences and, simultaneously, have all the required knowledge for formulating and implementing the desired policy programmes. Usually, two types of integrated planning procedures can be distinguished: horizontal and vertical integrated planning. The first refers to programmes and/or projects which attempt to integrate different factors in society (physical, socio-economic and environmental aspects) and takes place at various spatial levels (national, regional or local). The second approach refers to various sectorial issues (such as, housing, transportation and water supply) and links various governmental levels.

The success of these integrated planning procedures is dependent predominantly upon the institutional structures in which they are embedded. The structural and behavioural changes expected of these institutions would require emphasis on goal achievement rather than on rule-observation; substituting vertical with horizontal relationships and recognizing mutual interdependencies between different agencies of the bureaucracy. However, changes of the institutional structure are necessary but not sufficient for achieving effective integration. In addition, it is important to agree on common objectives to secure coherence and consistency in the planning process. The administrative structure alone may only be a facade which conceals substantial differences and conflicts. The actual institutional structure of integrated planning mechanisms and the extent of interpretation vary according to different environments. Developing countries with limited resources, in particular trained manpower, will only be able to introduce rudimentary forms of integrated planning. Thus, appropriate organizational structures are essential, at all levels of government and society, to facilitate integrated planning for they provide the basis for its continuity.

Data Collection

If integrated planning is considered as an evolving process, monitoring is essential to permit regular assessment to be made, to test the achievement of the objectives, update the recommended measures and policies and evaluate the priorities of implementation. This, however, requires a continuous flow of relevant data. In fact, the whole concept of integrated planning, as a technique, relies upon data, which includes not only data on ever-changing situations (for example, trends of environmental quality indicators or population trends), but also information about the effectiveness of prescribed policies. With respect to developing countries, which lack appropriate organizational structures and the availability of adequate data bases, more research efforts in the area of soft data analysis and planning methods are necessary and beneficial.

In conclusion, integrated planning is an attempt to cope with the complexity of the physical, socio-economic and environmental systems. Fully integrated planning is obviously the dream of a perfectionist and, as such, it contains the risk of becoming a cumbersome and slow process. It is an appropriate and useful concept for public policy making. But it needs to remain a flexible and adaptive approach to society's demands and not to become a dogmatic, over-ambitious attempt at comprehensiveness.

PARTICIPANTS

Ad Hoc Expert Meeting on
Integrated Physical, Socio-economic and Environmental Planning

Paris, 2-4 September 1981

Dr. Igor V. Chirgadze,
Deputy Chairman,
State Planning Committee,
Uzbek SSR, USSR

Professor Dr. Paul B. Downing,
Policy Sciences Programme,
Florida State University,
Florida, USA

Professor Dr. Bohdan Gruchman,
Academy of Economics,
Poznan, Poland

Professor Dr. Niles Hansen,
Department of Economics,
The University of Texas,
Austin, Texas, USA

Dr. K. Leonhardt,
Economic Commission for Europe (ECE),
Environment and Human Settlements Division

*Ms. Dorothy Nelkin,
Cornell University,
Ithaca, USA

Professor Dr. Peter Nijkamp,
Free University of Amsterdam,
The Netherlands

*Not present at the meeting

Mr. Francis Wasserman,
Research Fellow of SEDES,
Paris, France

Dr. Per Wramner,
Under Secretary of State,
Ministry of Housing and Physical Planning,
and Special Adviser on Environmental
Development,
Ministry of Foreign Affairs,
Stockholm, Sweden

Dr. Klaus Zimmermann,
International Institute of Environment
and Society,
Science Centre,
Berlin, Federal Republic of Germany

FROM UNITED NATIONS ENVIRONMENT PROGRAMME (UNEP)

Yusuf J. Ahmad,
Director for Special Assignments, UNEP

Professor Dr. Frank G. Müller,
Chief, Policies
and Applications Unit, UNEP

Mrs. Shadia Sawiris Schneider,
Consultant, UNEP